# DNA Replication - Epigenetic Mechanisms and Gene Therapy Applications

*Edited by Ziyad S. Haidar*

Published in London, United Kingdom

DNA Replication - Epigenetic Mechanisms and Gene Therapy Applications
http://dx.doi.org/10.5772/intechopen.104310
Edited by Ziyad S. Haidar

Contributors
Diriba Beyene Goonde, Seltene Abady, Sundru Manjulata Devi, Surya Chandra Rao Thumu, Shobha Rani Papanna, Ahmed Mohamed Nabil Helaly, Doaa Ghorab, Ziyad S. Haidar



First published in London, United Kingdom, 2023 by IntechOpen
IntechOpen is the global imprint of INTECHOPEN LIMITED, registered in England and Wales, registration number: 11086078, 5 Princes Gate Court, London, SW7 2QJ, United Kingdom

British Library Cataloguing-in-Publication Data
A catalogue record for this book is available from the British Library

Additional hard and PDF copies can be obtained from orders@intechopen.com

DNA Replication - Epigenetic Mechanisms and Gene Therapy Applications
Edited by Ziyad S. Haidar
p. cm.
Print ISBN 978-1-80356-875-1
Online ISBN 978-1-80356-876-8
eBook (PDF) ISBN 978-1-80356-877-5

# Meet the editor

Ziyad S. Haidar is a Full Professor of Biomaterials, Nanotechnology and Tissue Engineering and the scientific director of Facultad de Odontología, Universidad de los Andes (UANDES), Santiago, Chile. He is the founder and head of the Biomaterials, Pharmaceutical Delivery and Cranio-Maxillo-Facial Bio-Engineering Laboratory/Research Group (BioMAT'X Chile-HAiDAR R&D&I), head of innovation at Centro de Investigación e Innovación Biomédica (CiiB), and a faculty member in the dental sciences and biomedicine doctoral programs at the Facultad de Odontología and Facultad de Medicina, UANDES, respectively. He is also a visiting professor at several institutions, including the Division of Cranio-Maxillo-Facial Surgery, Universidad de La Frontera (UFRO), Temuco, Chile. Dr. Haidar is an oral and cranio-maxillo-facial surgeon with a Ph.D. in Nanobiomaterials, Pharmaceuticals and Bioengineering from McGill University, Canada. He completed a post-doctoral training residency at the Montréal Shriners Hospital, MUHC, Montréal, QC, Canada.

# Contents

# Preface

As a professional duty, editing a book demands considerable effort. A book editor is responsible for assisting contributors in reaching their full potential and ensuring that the final product is presented to readers with professionalism and accuracy. During 2020–2021, I served as editor for the IntechOpen book entitled *Biomechanics and Functional Tissue Engineering*, a useful resource for scientists, researchers, engineers, and clinical practitioners involved in investigating and developing bio-engineering solutions for improving patient quality of life. The content focused on applied and functional bio-dental tissue engineering, drug/gene delivery (controlled and metered systems) and cell (stem cell) therapy, bio-scaffolds, image-guided and image-assisted surgery, nano-dentistry, regenerative medicine, and the involved biomechanics thereof. In this project, better innovative solutions and strategies tackling acute and chronic problems and conditions of the oro-dental and cranio-maxillo-facial complex and beyond were presented. Moving and/or implementing benchtop research, development, and innovation (R&D&I) to market and/or clinical practice (i.e., translation) is a lengthy and tedious process, often requiring years. Consequently, a second IntechOpen book completed during 2021–2022 aimed to address the fields of biomimetics, bio-design, and bio-inspiration, all of which are relatively new areas of science and R&D&I where researchers look to mimic (or rather emulate) aspects of nature to help solve human problems. This project presented biomimicry in a simplified and practical manner. *Biomimetics - Bridging the Gap* presented the biomimicry processes as either problem- or solution-based. Today, biomimicry is considered the most advanced process of applying cellular and biological principles that underlie morphology, structures, and functionality of physiological entities to design and formulate or fabricate human-made efficient solutions for persistent challenges. In the book, we wondered what new lessons can be learned to live better?

Understanding DNA replication, methylation, epigenetics, and gene therapy is integral to designing and developing novel treatments for various challenging diseases. By studying such natural mechanisms and sophisticated processes, researchers can gain insights into how to, for example, modulate gene expression in a controlled and targeted manner, providing opportunities for developing novel drug delivery systems and gene therapies. Furthermore, biomimicry was suggested to have the capability to inspire the development of novel gene editing tools that, in turn, would mimic natural DNA repair mechanisms. For instance, consider the CRISPR-Cas9 system, a revolutionary gene editing tool, inspired by the bacterial immune system. Consequently, the present book *DNA Replication – Epigenetic Mechanisms and Gene Therapy Applications* revisits the basic fundamentals and the latest trends relevant to DNA replication research, including the involved molecular mechanisms of initiation and termination, functional cross-talk and/or association to the genome, DNA recombination and repair, and analytical methods, presenting recent and emerging applications and discussing the future directions of the field. This book is a concise yet thorough resource that brings together and synthesizes contributions from active and prominent researchers in the field, with chapters addressing such topics as basic molecular biology of the cell, cell growth and division, DNA isolation and duplication, DNA damage response, epigenetics, DNA methylation, and gene silencing. It also reviews the most recent and advanced (modern) analytical and computational tools such as drug delivery systems and genomic imprinting and their relevance to human disease. There is also a section dedicated to dentistry and oral health, highlighting the revolutionary potential and use of DNA replication in a range of clinical and surgical treatments, procedures, and interventions, such as dental pulp inflammation repair, the fate of dental mesenchymal stromal/stem cells, gingival and periodontal therapy, augmented dental implantology, and whole tooth bioengineering.

This book is organized into three sections.

Section 1, "Introduction to DNA Replication, Methylation, Epigenetics and Gene Therapy", provides simplified definitions for DNA and RNA methylation and the mechanisms of histone modification and presents a variety of epigenetic modifications that can lead to anti-tumor drug resistance, amongst other potential and pertinent applications. By attempting to unravel the complexities of DNA replication, epigenetics, and gene therapy, the section also explores how targeting epigenetic modifiers can reverse drug resistance. It also discusses the wide-ranging potential applications of DNA replication, epigenetics, and gene therapy in the field of oral health care, dentistry, and cranio-maxillo-facial surgery.

Section 2, "DNA Replication Mechanisms and Epigenetics", dives deeper into the processes of DNA and RNA methylation and the mechanisms of histone modification. It presents a variety of modern pharmacotherapeutic approaches and targeted epigenetic modifications that can lead to controlled anti-tumor drug resistance. It provides valuable insights into controlled delivery and how epigenetic therapy and targeting epigenetic modifiers can reverse drug resistance in contemporary and future oncology.

Section 3, "Gene Therapy Applications", amalgamates the previous sections and chapters by providing expert perspectives and applied examples of gene therapy through discussion of specific and clinically relevant scenarios. Topoisomerase 1 (TOP1) as a therapeutic target and its relationship with colorectal carcinoma is a fine example. Colorectal carcinoma, commonly known as colorectal cancer, is cancer that develops in the colon or rectum. It is the third most common cancer globally and has various underlying genetic and molecular alterations contributing to its development and progression. TOP1 is an enzyme involved in the regulation of DNA topology, specifically in the relaxation of supercoiled DNA during processes like DNA replication and transcription. It accomplishes this by introducing transient single-strand breaks in DNA, allowing the DNA strands to rotate and relieve tension. More importantly, TOP1 has been identified as a potential therapeutic target in colorectal carcinoma. Several chemo-therapeutic agents, such as irinotecan, work by inhibiting TOP1 activity. These drugs stabilize the cleavable complex formed by TOP1 on DNA, leading to DNA damage and cell death. Irinotecan is commonly used in the treatment of colorectal cancer, particularly in advanced or metastatic stages. Alterations in the TOP1 gene itself have been observed in colorectal carcinoma. These mutations can affect TOP1 function, leading to changes in DNA topology regulation. Furthermore, TOP1 mutations are associated with an increased risk of developing colorectal cancer and thereby can influence response to TOP1-targeted therapies. Studies have examined the expression levels of TOP1 in colorectal carcinoma and its association with patient prognosis. High levels of TOP1 expression have been linked to poorer outcomes in terms of disease progression and survival rates. This suggests that TOP1 expression could serve as a potential prognostic marker in colorectal cancer. This section highlights that resistance to TOP1-targeted therapies, such as irinotecan, can also develop in colorectal carcinoma patients. Mechanisms of resistance include alterations in TOP1 expression or function, DNA repair pathways, drug efflux pumps, and other cellular processes. Understanding these resistance mechanisms is deemed crucial for developing novel methods, strategies, and modalities to help overcome drug resistance and thus improve treatment outcomes.

The integration of DNA replication, epigenetics, and gene therapy discussed in this book has the potential to revolutionize medicine and improve patient outcomes, survival, and quality of life. As we move towards personalized and precise medicine, it is important to remember that the genetic material is identical in every cell, while epigenetics is highly variable within different cells and tissues of an organism and is affected by aging and environmental factors. As such, identifying and validating such novel epigenetic modifications associated with cancer chemoresistance in clinical studies cannot be underestimated. As researchers continue to make breakthroughs, the translation of these discoveries from benchtop to bedside will be crucial for delivering effective treatments to our patients. On a final note, the possible and potential applications of DNA replication, epigenetics, and gene therapy in medicine, oncology, and oral and dental health care, amongst other related fields are vast, and with ongoing R&D&I, state-of-the-art literature, and cutting-edge technologies driving the field forward, it is highly likely to yield many exciting new findings, tools, and applications in the years to come.

*"Utilizing the scientific knowledge gained from genetics and epigenetics can strengthen the effectiveness of innovation, novel therapies and thereby alleviate human suffering."*

**Dr. Ziyad S. Haidar,**
Professor,
BioMAT'X (HAiDAR R&D&I/I+D+i LAB),
Centro de Investigación e Innovación Biomédica/
Biomedical Research and Innovation Center (CiiB),
Faculties of Dentistry and Medicine (Cross-Appointment),
Universidad de los Andes,
Las Condes, Santiago de Chile, Chile

Section 1

# Introduction to DNA Replication, Methylation, Epigenetics and Gene Therapy

Chapter 1

# Introductory Chapter: Unraveling the Complexities of DNA Replication, Epigenetics, and Gene Therapy Applications

*Ziyad S. Haidar*

## 1. Introduction

"DNA Replication: *Epigenetic Mechanism and Gene Therapy Applications*" is primarily conceived to present the latest trends of the DNA replication research field via presenting the basic fundamentals, involving molecular mechanisms of initiation and termination, functional cross-talk and/or association to the genome, DNA recombination and repair, analytical methods, recent applications and discussing the future directions of the area; a highly-exciting and -timely book to the interested reader, designed to provide a dedicated focus on the clinical contexts and translational processes. Hence, this book aims to bring together and synthesize contributions from active and prominent researchers, with the chapters ranging from the basic molecular biology of the cell, cell growth and division, DNA isolation, duplication, and DNA damage response, to epigenetics, DNA methylation, and gene silencing, to a review of the most recent and advanced (modern) analytical and computational tools, to the relevance of all this to human disease, in general, drug delivery and genomic imprinting in specific and last but not least, a dedicated section to dentistry and oral health, highlighting the revolutionary potential and use of DNA replication, via an applied demonstration in a range of clinical treatments, procedures, and interventions, such as dental pulp inflammation repair, the fate of the dental cells, periodontal therapy, augmented dental implantology, and whole tooth bioengineering. Henceforth, "DNA Replication: *Epigenetic Mechanism and Gene Therapy Applications*" is a comprehensive guide to the latest research and advancements in the field of genetics. The book provides an in-depth exploration of DNA replication, epigenetics, and gene therapy applications, offering a detailed understanding of the mechanisms that govern our genetic makeup and how they can be manipulated to improve human health. The book is a suitable resource for students, researchers, and medical professionals looking to expand their knowledge of genetics and explore new avenues for therapeutic interventions. With contributions from leading experts, it is a must-read for anyone interested in understanding the fundamental processes that underlie our genetic code, in a simplified manner. For instance, the introductory chapter aims to define DNA and RNA methylation, the mechanisms of histone modification, and presents a variety of epigenetic modifications which can lead to anti-tumor drug resistance. It also explores how targeting epigenetic modifiers can reverse drug resistance.



IntechOpen

The fields of DNA replication, epigenetics, and gene therapy are all complex and rapidly evolving, with many exciting developments and challenges ahead [1, 2]. Briefly, DNA replication involves the intricate molecular mechanisms that ensure accurate duplication of genetic material, and hence, understanding these processes is crucial for various fields, including genetics, cancer research, and drug discovery [3, 4]. Epigenetics, on the other hand, focuses on the regulation of gene expression through modifications of DNA and its associated proteins and has emerged as a vital area of research with implications for aging, development, and disease [3–7]. Last but not least, gene therapy aims to correct or modify genetic material in order to treat or prevent disease, has the potential to revolutionize medicine [8, 9], and nonetheless also poses significant challenges related to gene delivery, bio-efficacy, and -safety [10–12]. It can be projected that in the future, the integration of these sub-fields will likely continue to drive the advances in personalized medicine, with gene therapy and epigenetic modifications being used to target specific diseases and individual patients [10, 12]. Additionally, ongoing research in DNA replication and repair mechanisms will likely lead to the development of new therapies and drugs for cancer and other genetic diseases [13–15]. Henceforth, it is worth mentioning that while there are still many unknowns and complexities to be navigated, the potential benefits of this research and innovation area are immense and will undoubtedly play a critical role in shaping the future of medicine and biotechnology [1, 3, 6, 10].

## 2. DNA replication

DNA replication is the process by which a cell makes an identical copy of its DNA [1, 2]. It is a crucial process that occurs in all living organisms and is necessary for cellular division, mitosis, and the transmission of genetic information from one generation to the next [2, 4]. Basically, during DNA replication, the double-stranded DNA molecule is unwound and separated into two strands. Each strand then serves as a template for the synthesis of a new complementary strand by the attachment of nucleotides in a specific order, according to the base-pairing rules (A-T, C-G). The end result is two identical DNA molecules, each consisting of one original and one new strand [1–6]. DNA replication (**Figure 1**) is a complex and highly regulated process crucial for the accurate transmission of genetic information from one generation of cells to the next. It involves the duplication of the entire DNA molecule in a semi-conservative manner, meaning that each new DNA molecule holds one original strand and one newly synthesized strand [1, 4, 6].

## 3. Epigenetics

Epigenetics is the study of changes in gene expression that occur without any alteration in the DNA sequence itself [7]. These changes can be heritable and reversible and are caused by modifications to the DNA molecule and its associated proteins, which regulate gene expression. Examples of epigenetic modifications include DNA methylation, histone modification, and non-coding RNA molecules. Epigenetic changes can influence gene expression by either activating or silencing genes, and they play a crucial role in various biological processes such as development, aging, and disease. Likewise, epigenetics is a rapidly growing field that has the potential to revolutionize our understanding of gene regulation with effects on many areas of bio-medicine, such as cancer, neurological disorders, personalized-medicine and -dentistry.

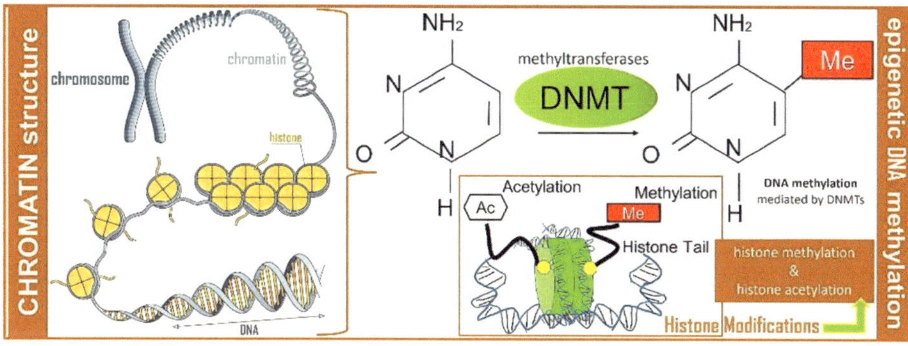


**Figure 1.**
*DNA methylation and replication and some of the involved enzymes and proteins. DNA methylation is a biochemical process that involves the addition of a methyl group (-CH3) to the DNA molecule. It is a fundamental mechanism of epigenetic regulation, which plays a crucial role in gene expression and cellular differentiation. DNA methylation is a dynamic process and primarily occurs at cytosine residues within CpG dinucleotides, where a cytosine is followed by a guanine in the DNA sequence. It involves: DNA methyltransferases (DNMTs): DNA methylation is catalyzed by a group of enzymes called DNA methyltransferases. The two main DNMTs involved in establishing and maintaining DNA methylation patterns are DNMT3A and DNMT3B, responsible for de novo methylation, and DNMT1, which maintains existing methylation patterns during DNA replication; CpG Islands: CpG islands are regions of DNA that contain a high frequency of CpG dinucleotides. These regions are often found near the promoter regions of genes and are involved in the regulation of gene expression. CpG islands can be either unmethylated, associated with active gene expression, or methylated, associated with gene silencing; methylation patterns: DNA methylation patterns can vary across different cell types, tissues, and developmental stages. They are heritable during cell division and can be influenced by environmental factors and cellular signaling pathways. Methylation patterns can undergo dynamic changes throughout life, playing a crucial role in cellular processes such as embryonic development, X-chromosome inactivation, genomic imprinting, and gene regulation; methylation and gene regulation: DNA methylation typically represses gene expression by interfering with the binding of transcription factors and other regulatory proteins to the DNA sequence. Methylated CpG sites can recruit proteins that modify chromatin structure, leading to a condensed chromatin state (heterochromatin) and preventing gene activation. However, DNA methylation can also be associated with gene activation in certain contexts, such as gene bodies; and DNA demethylation: In addition to DNA methylation, DNA demethylation processes exist to remove or alter DNA methylation patterns. Passive DNA demethylation occurs during DNA replication when newly synthesized DNA strands lack methyl groups, leading to a gradual loss of DNA methylation over successive cell divisions. Active DNA demethylation involves enzymatic processes that actively remove or modify methyl groups, including 10–11 translocation (TET) enzymes. Please note that DNA methylation can be influenced by various factors, including environmental cues, hormonal signals, and disease states. The precise patterns of DNA methylation are crucial for normal development and cellular function, and aberrant DNA methylation can be associated with various diseases, including cancer, neurological disorders, and imprinting disorders. On the other hand, the DNA replication process is tightly regulated to ensure accuracy and efficiency. Various checkpoints and control mechanisms exist to monitor the integrity of the DNA and ensure that any errors or damage are corrected before the replicated DNA is passed on to the next generation of cells. During DNA replication, several enzymes and proteins work together to coordinate and carry out the process. Some of the key players involved are helicase: This enzyme is responsible for unwinding the double helix structure of the DNA molecule, separating the two strands and creating a replication fork; DNA polymerase: These enzymes are responsible for synthesizing new DNA strands by adding complementary nucleotides to the existing template strands. There are several types of DNA polymerases involved in different stages of replication, including the main replicative polymerases (DNA polymerase alpha, delta, and epsilon) and the DNA repair polymerases; primase: Primase is an enzyme that synthesizes short RNA primers on the DNA template. These primers provide a starting point for DNA polymerase to initiate DNA synthesis; single-Strand binding proteins (SSBs): These proteins bind to the separated DNA strands and stabilize them, preventing them from re-annealing or being degraded by nucleases; DNA ligase: DNA ligase is responsible for sealing the gaps between the newly synthesized DNA fragments, known as Okazaki fragments, on the lagging strand; topoisomerases: These enzymes relieve the tension and strain that builds up ahead of the replication fork by cutting and rejoining the DNA strands; and DNA proofreading and repair proteins: These proteins monitor the replication process and check for errors. If any errors are detected, they can remove the mismatched nucleotides and replace them with the correct ones. Remember that the precise patterns of DNA methylation are crucial for normal development and cellular function, and aberrant DNA methylation can be associated with various diseases, including cancer, neurological disorders, amongst others.*

## 4. Gene therapy

Gene therapy can be defined as a *biomedical* approach that involves the introduction, removal, and/or alteration of genetic material within the cells of an individual to either treat or prevent disease [8]. Herein, this genetic material can be in the form of DNA, RNA, or proteins and can be delivered to cells using various methods such as viral vectors, liposomes, or nanoparticles (and gene editing tools). The aim of gene therapy is to correct or replace faulty genes, modify gene expression, or introduce new *functional* genes to cells, with the goal of curing or alleviating genetic disorders, as well as some non-genetic diseases such as cancer and infectious diseases [9, 10]. The process typically involves: [1] *Vector Selection* where a suitable vector, often a modified virus or a plasmid, is chosen to deliver the therapeutic gene into the target cells. The vector acts as a vehicle to transport the desired genetic material; [2] *Gene Insertion* where the therapeutic gene or the modified gene is inserted into the chosen vector, which is then introduced into the target cells. Various techniques, such as viral-mediated gene transfer or non-viral methods like electroporation or lipofection, can be used for gene delivery; [3] *Cellular Uptake* where the target cells, depending on the specific strategy, take up the vector carrying the therapeutic gene. Once inside the cells, the vector releases the therapeutic gene; and [4] *Transcription and Translation*: where the introduced therapeutic gene is transcribed into messenger RNA (mRNA) by the cellular machinery, including enzymes involved in DNA replication. The mRNA is then translated into a functional protein. Remember that while gene therapy is not directly related to the process of DNA replication, it does rely on the understanding of DNA replication and the cellular machinery involved in DNA synthesis. Henceforth, gene therapy is an exciting and rapidly evolving field that holds great promise for the treatment of a wide range of diseases [11, 13]. Yet, as with any new technology, it poses some challenges related to the safety, specificity, efficacy, and delivery of genetic material to target cells (gene therapy can lead to activating oncogenes which can cause cancer and the use of viral vectors can also trigger an immune response in the body leading to inflammation amongst other adverse effects). **Note**: The successful integration and expression of the therapeutic gene rely on the cellular processes involved in DNA replication, including the availability of DNA polymerases, DNA repair mechanisms, and other proteins that facilitate DNA synthesis and gene expression [4, 7, 8, 10].

## 5. Perspective

Today, DNA replication, epigenetics, and gene therapy are at the forefront of cutting-edge research and innovation with many recent advances and promising developments [12, 14, 15]. In laboratories, researchers are working to unravel the complexities of DNA replication, with a particular focus on the mechanisms of initiation and termination, functional cross-talk with the genome, and DNA recombination and repair, for example. Further, new analytical methods, such as single-cell sequencing and CRISPR-based gene editing, are also being developed to enhance our understanding of DNA replication and its role in human health and disease. In the sub-field of epigenetics, recent studies have revealed the potential of epigenetic modifications, such as DNA methylation, histone modification, and non-coding RNA molecules, to be used as therapeutic targets for a range of diseases [15]. For example, recent research has shown that targeting epigenetic modifiers can reverse drug resistance in certain cancers, while other studies have demonstrated the potential of epigenetic therapies for neurological disorders such as Alzheimer's disease. Gene therapy, meanwhile, is also advancing rapidly from bench-top to bed-side, with a growing number of clinical trials and approved treatments. Recent successes include the development of gene therapies for genetic disorders such as spinal muscular atrophy and inherited retinal disease, as well as for non-genetic diseases such as cancer and infectious diseases. Patents are being filed for new gene therapies, such as those utilizing CRISPR-Cas9 gene editing technology, with an increasing interest in the use of gene therapy for personalizing health care.

In the field of oral health care, dentistry, and cranio-maxillo-facial surgery, in particular, there are many potential applications of DNA replication, epigenetics, and gene therapy [8–11]. Here are a few examples (**Figure 2**): *Dental pulp inflammation repair*: Gene therapy has been investigated as a potential treatment for dental pulp inflammation, which can lead to tooth loss. Researchers have explored the use of gene therapy to deliver anti-inflammatory cytokines to the inflamed pulp tissue, which could promote healing and prevent further damage to the tooth. *Fate of dental-derived mesenchymal stem/stromal cells*: DNA replication and epigenetic modifications play a critical role in the development and differentiation of dental cells. Understanding these processes could help researchers to develop new therapies for conditions such as tooth loss and gum/periodontal disease. *Periodontal therapy*: Epigenetic modifications, such as DNA methylation and histone modifications, have been shown to play a role

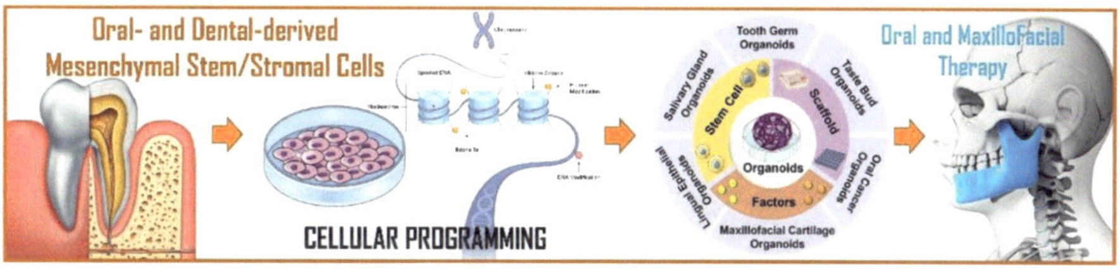


**Figure 2.**
*DNA replication, epigenetics, and gene therapy in Oral, dental, and Cranio-Maxillo-facial applications. While applications do hold promise, please note that they are still under active research and development and innovation (R&D&I). The translation of concepts into clinical practice requires further investigation, ensuring safety, efficacy, and long-term outcomes. Nonetheless, advancements in DNA replication, epigenetics, and gene therapy have the potential to significantly impact Oral health care and cranio-maxillo-facial surgery in the future. Some examples include genetic disorders: Gene therapy holds promise for treating genetic disorders that affect oral health, such as orofacial clefts (e.g., cleft lip and palate) and genetic syndromes associated with dental abnormalities. By introducing functional genes or correcting genetic mutations, gene therapy could potentially address the underlying genetic causes of these conditions; periodontal disease: Epigenetic modifications, including DNA methylation and histone modifications, have been implicated in the development and progression of periodontal disease. Understanding the epigenetic mechanisms involved could lead to the identification of novel therapeutic targets for preventing or treating periodontal disease; salivary diagnostics: Saliva contains DNA from oral tissues, and the analysis of saliva-derived DNA can provide valuable diagnostic information. DNA replication techniques, such as polymerase chain reaction (PCR), are commonly used in salivary diagnostics to detect and quantify specific pathogens or genetic markers associated with oral diseases; tissue engineering and regenerative medicine: DNA replication plays a crucial role in cell proliferation and tissue regeneration. In dentistry and cranio-maxillo-facial surgery, tissue engineering approaches aim to regenerate damaged or lost oral and facial tissues, such as bone, cartilage, and teeth. Gene therapy can be used to enhance tissue regeneration by introducing genes that promote cell growth, angiogenesis, or extracellular matrix synthesis; pain management: Epigenetic modifications can influence pain perception and sensitivity. Understanding the epigenetic mechanisms involved in chronic orofacial pain conditions could lead to the development of targeted therapies for pain management in dentistry; and oral cancer: Epigenetic alterations, including DNA methylation and histone modifications, play a critical role in the development and progression of oral cancer. Epigenetic therapies, such as DNA demethylating agents and histone deacetylase inhibitors, are being investigated as potential therapy options for oral cancer (and others).*

in the development and progression of periodontal disease. Researchers are exploring the potential of epigenetic therapies to target these modifications and prevent or treat periodontal disease. *Augmented dental implantology*: Gene therapy has been investigated as a potential approach to enhance the integration of dental titanium implants with the surrounding hard/soft tissue. Researchers have explored the use of gene therapy to deliver growth factors and other molecules to promote osteogenesis and growth of other tissues around the implant. *Whole tooth bioengineering*: Researchers are exploring the use of gene therapy and tissue engineering techniques to develop fully functional teeth that can be implanted in patients who have lost their natural teeth. This could revolutionize oral and dental care, particularly for patients who cannot opt for traditional dental implants due to insufficient bone (quantity/quality).

## 6. Closing remarks

In conclusion, the integration of DNA replication, epigenetics, and gene therapy has the potential to revolutionize medicine and improve patient outcomes [3, 14, 15]. As researchers continue to make breakthroughs, the translation of these discoveries from bench-top to bed–/chair-side will be crucial for delivering effective treatments to our patients. Indeed, the possible applications of DNA replication, epigenetics, and gene therapy in medicine, oncology, oral and dental health care amongst other related fields are vast, and with ongoing research, development and innovation, state-of-the-art literature, and cutting-edge technologies driving the field forward, it is likely to yield many exciting new discoveries and applications in the years to come.

## Conflict of interest

The author declares no conflict of interest.

## Author details

Ziyad S. Haidar
BioMAT'X (HAiDAR R&D&I LAB) I+D+i Labratory, Biomedical Research and Innovation Center/Centro de Investigación e Innovación Biomédica (CiiB), Faculties of Medicine and Dentistry, Universidad de los Andes, Santiago, Chile

*Address all correspondence to: zhaidar@uandes.cl; zhaidar78@gmail.com

## References

[1] Ekundayo B, Bleichert F. Origins of DNA replication. PLoS Genetics. 2019;**15**(9):e1008320

[2] Nordman J, Orr-Weaver TL. Regulation of DNA replication during development. Development. 2012;**139**(3):455-464

[3] Peixoto P, Cartron PF, Serandour AA, Hervouet E. From 1957 to nowadays: A brief history of epigenetics. International Journal of Molecular Sciences. 2020;**21**(20):7571

[4] Song H-Y, Shen R, Mahasin H, Guo Y-N, Wang D-G. DNA replication: Mechanisms and therapeutic interventions for diseases. MedComm. 2023;**4**:e210

[5] Delgado-Morales R, Agis-Balboa RC, Esteller M, Berdasco M. Epigenetic mechanisms during ageing and neurogenesis as novel therapeutic avenues in human brain disorders. Clinical Epigenetics. 2017;**9**:67

[6] Muñoz S, Méndez J. DNA replication stress: From molecular mechanisms to human disease. Chromosoma. 2017;**126**(1):1-15

[7] Ganesan A, Arimondo PB, Rots MG, Jeronimo C, Berdasco M. The timeline of epigenetic drug discovery: From reality to dreams. Clinical Epigenetics. 2019;**11**(1):174

[8] Scheller EL, Villa-Diaz LG, Krebsbach PH. Gene therapy: Implications for craniofacial regeneration. The Journal of Craniofacial Surgery. 2012;**23**(1):333-337

[9] Gupta K, Singh S, Garg KN. Gene therapy in dentistry: Tool of genetic engineering. Revisited. Archives of Oral Biology. 2015;**60**(3):439-446

[10] Siddique N, Raza H, Ahmed S, Khurshid Z, Zafar MS. Gene therapy: A paradigm shift in dentistry. Genes (Basel). 2016;**7**(11):98

[11] Fliefel R, Kühnisch J, Ehrenfeld M, Otto S. Gene therapy for bone defects in Oral and maxillofacial surgery: A systematic review and meta-analysis of animal studies. Stem Cells and Development. 2017;**26**(4):215-230

[12] Wilhelm T, Said M, Naim V. DNA replication stress and chromosomal instability: Dangerous liaisons. Genes (Basel). 2020;**11**(6):642

[13] Larsson L. Current concepts of epigenetics and its role in periodontitis. Current Oral Health Reports. 2017;**4**(4):286-293

[14] Zeng B, Liu G, Huang J. DNA methylation and histone modification in dental-derived mesenchymal stem cells. Stem Cell Reviews and Reports. 2022;**18**(8):2797-2816

[15] Zhang H, Fu H, Fang H, Deng Q, Huang H, Hou D, et al. Epigenetic regulation of methylation in determining the fate of dental mesenchymal stem cells. Stem Cells International. 2022:5015856. DOI: 10.1155/2022/5015856. PMID: 36187229; PMCID: PMC9522499

Section 2

# DNA Replication Mechanisms and Epigenetics

Chapter 2

# Anti-Tumor Drug Resistance and Modern Oncologic Pharmaco-Therapy: RNA and DNA Methylation, Mechanisms and Histone Modification, Epigenetic Regulation and Targeting Epigenetic Modifiers in Contemporary Cancer Therapy

*Ziyad S. Haidar*


## Abstract

Metastasis, the spread of cancer cells from the primary tumor to the surrounding tissues and to distant organs, is one (and perhaps the primary) of the major causes of cancer-related death (or cancer morbidity and mortality). Indeed, it is estimated that metastasis is responsible for about 90% of cancer deaths. The major factors contributing to the metastasis of cancer cells are epithelial-mesenchymal transition (EMT) and cancer stem cells (CSCs). Herein, the cancer cells must detach from the primary tumor, intravasate into the circulatory and lymphatic systems, evade immune attack, extravasate at distant capillary beds, and invade and proliferate in distant organs. Accruing evidence suggests that the malfunction of epigenetic regulation in the functioning of a gene is directly related to the generation of tumors and cancer. Henceforth, the potential and capacity to change or re-program the epigenetic landscape within the epigenome of cancer is possibly the most promising and pursued targeted therapy, nowadays. Such would lead to reversing drug resistance and so, new therapeutic modalities. Indeed, contemporary oncologic pharmaco-therapy for cancer has and continues to undergo remarkable changes; especially lately, in terms of the introduction of effective cancer-specific molecular-targeted therapeutic agents. This introductory chapter to the book titled: "DNA Replication – Mechanisms, Epigenetics, and Gene Therapy Applications" discusses DNA and RNA methylation, the mechanisms of histone modification, and presents a variety of epigenetic modifications which can lead to anti-tumor drug resistance. It also explores how targeting epigenetic modifiers can reverse drug resistance.

**Keywords:** DNA replication, epigenetics, gene therapy, RNA methylation, histone modification, drug resistance, oncology, therapy

## 1. Introduction

Clinically, selecting the ideal anti-cancer therapy might consider combining epigenetic-related drugs. Indeed, epigenetic regulation and epigenetic changes, during treatment have been reported as one of the correlating mechanisms for anti-tumor drug resistance. This is due to the fact that during treatment, the development of drug-resistant tumors continues to pose a critical challenge in oncology; severely increasing the mortality rate, worldwide. This resistance can be categorized as either *de novo* or acquired depending on whether the resistance is inherent or has been acquired due to continuous drug administration [1]. Mostly, the research on cancer drug resistance has focused on the genetic aspect of the disease but the importance of epigenetic regulation is coming to light. Epigenetics events have always been important in the progression of cancer. Two very important epigenetic events that affect the expression of genes are methylation and acetylation which activates the oncogenes and reduces the tumor suppressor genes which lead to cancer drug resistance. Herein, both, ovarian and breast cancers are tumors whose epigenetic basis has been studied in detail [2]. Tyrosine kinase inhibitors have an important role in the therapy of cancer. The effectiveness of (TKIs) is good for both solid and liquid cancer. This is used to reduce the oncogenic activity with the epidermal growth factor receptor (EGFR), as well as tyrosine kinase receptor has developed resistance against EGFR-TKIs [3]. The role of demethylases in resistance against TKIs upregulation of these demethylases leads to resistance against TKIs. The events of epigenetics are not limited to solid tumors, a common childhood malignancy (lymphoblastic leukemia) where the changes in genetics are not enough to prove the increased relapse and chemo-resistance [4]. The less studied portion of the cell is the *centrosome* (archaically cyto-center) in the context of epigenetic regulation of drug resistance. The centrosome organelle (microtubule organizing center) has a very important role during the division of the cell in which the centrosome distributes cellular components within daughter cells. Any abnormal function of the centrosome in epigenetic events can degrade the integrity of the cells, resulting in genetic instability [5]. Consequently, the main issue in anti-cancer therapeutics is the development of resistance which has become the biggest problem in cancer survival rates. Indeed, resistance to cancer therapy can develop in a multitude of ways which also includes alteration in epigenetics in cancer cells [6]. There are many ways by which cancer cells reconstruct their epigenomics landscape so that they can resist anti-cancer treatment. To tackle the effects of chemo-resistance, there are various modifiers that are used, including histone deacetylase inhibitors, DNA hypomethylating agents, and histone demethylase inhibitors. A modifier has an average success, whether when used alone or in combination, yet the best result, to date, was achieved when modifiers were used with some of the conventional or traditional anti-cancer therapeutics [7]. Herein, pharmaco-therapeutics that modify epigenomes succeed in weakening cancer cells via different processes, such as restoring cell cycle control, damaging pro-survival signaling, preventing DNA damage, reparation, and/or suppression of the immune system, to mention a few [8].

## 2. Cancer epigenetics: epigenetic mechanisms, regulation, and modification

As aforementioned, abnormal epigenetic modifications in specific oncogenes and tumor suppressors genes can result in un-controlled cell growth and division. Indeed, alterations in epigenetic modifications in cancer regulate various cellular responses, including cell proliferation, apoptosis, invasion, and senescence. Through DNA methylation, histone modification, chromatin remodeling, and noncoding RNA regulation, epigenetics play an important role in tumorigenesis. Nevertheless, it is worth-mentioning that abnormal epigenetic modifications in regions of DNA outside of genes, alongside other environmental factors, can also lead to or result in cancer. Numerous studies in the literature studied and continue to investigate the different landscapes of the genome from oncogene-driven signaling to all the mutation spectrum in all types of cancer sub-types [9]. It can be stated that the epigenetic modification can be categorized into three sub-types (carcinogenic mechanisms): DNA methylation, RNA methylation, and modification of histones and non-coding RNAs [10]. Those will be discussed in the next pages.

### 2.1 DNA methylation

Briefly, alterations in DNA methylation are common in development as well as in a variety of tumors. For example, if a gene necessary for DNA repair is hyper-methylated, this would result in deficient DNA repair and lead to an accumulation of DNA damage. An Increase in DNA damage will then cause increased errors during DNA synthesis, therefore, leading to mutations that can or tend to give rise to cancer. Hence, the methylation of DNA is one of the most studied mechanisms in epigenetics which usually occurs in CpG islands which is located at 5′ promoter region studies in more than 50% of human genes, this usually displays function in diseases and development which also include embryonic development, X chromosome inactivation, epigenetics reprogramming, genomic imprinting, and the establishment of cell identity [11]. Generally, it shows gene silencing with the addition of the groups of methyl from S-adenosylmethionine also known as (SAM) to five positions (cytosine pyrimidine ring) [12]. Now, this 5-methylcytosine structure can either restrict access of many transcriptional factors to the sites of binding in DNA, or it can take methyl binding domain proteins with histone modification to re-configure the chromatins, thus they can show expression of genes [13]. Generally, there are three DNA methyltransferases, and are known as DNMT1, DNMT3b, DNMT3a; which help in catalyzing the methylation of the DNA (**Figure 1**) [10].

#### *2.1.1 DNA methylation processes: de novo-, hyper-methylation, and demethylation*

DNA methylation regulates gene expression via recruiting proteins involved in gene repression or by inhibiting the binding of transcription factor(s) to the DNA. During development, the pattern of DNA methylation in the genome changes as a result of a dynamic process involving both *de novo* DNA methylation and de-methylation. If hyper-methylation occurs in DNA, the methyl group will attach to the cytosine at Carbon 5 and is catalyzed by the DNA methyl transferase, also known as (DNMTs), which also includes DNMT1, DNMT3b, DNMT3a. This methylation step of cytosine occurs at the CpG dinucleotides, and DNMT1 is highly responsible for maintaining the methylation by sending the methyl group to DNA strands which

| Nucleoside Analogue | Non-Nucleoside Analogue | Anti-Sense | Ref |
|---|---|---|---|
| Zebularine | RG 108 Procaine | MG98 | (14) |
| 5-Aza-2'-deoxycytidibe (Decitabine/ Dacogen) | RG 108 | | |
| 5'-Azacytidine (Vidaza) | Procainamide | | |

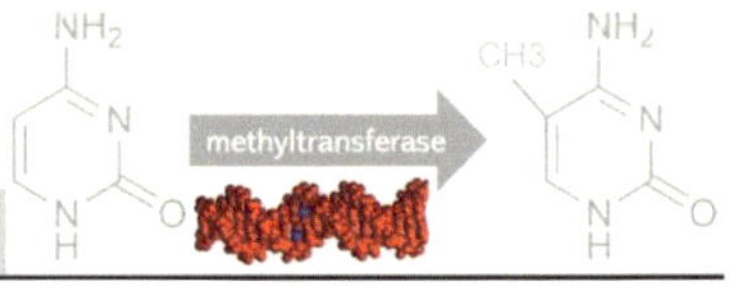


**Figure 1.**
*DNA methylation and regulators thereof. DNA methylation is a chemical modification process that involves the addition of a methyl group to the cytosine base of DNA molecules. It is an essential epigenetic mechanism that regulates gene expression and plays a crucial role in various biological processes. The process of DNA methylation is carried out by a group of enzymes called DNA methyltransferases (DNMTs), which are themselves regulated by various factors such as histone modifications, transcription factors, and non-coding RNAs. Regulators of DNA methylation are factors that control the activity and expression of DNMTs and thus modulate the level of DNA methylation. These regulators can be classified into two main categories: (1) positive regulators, which enhance the activity of DNMTs and increase DNA methylation levels, and (2) negative regulators, which inhibit the activity of DNMTs and decrease DNA methylation levels. Examples of positive regulators of DNA methylation include histone methyltransferases, which modify histone proteins and create a favorable chromatin environment for DNMTs to function, and certain transcription factors that recruit DNMTs to specific genomic regions. Negative regulators include DNA demethylases, which actively remove methyl groups from DNA, and some non-coding RNAs that can inhibit DNMT expression or activity. Overall, these regulators play a critical role in maintaining the appropriate DNA methylation patterns in cells and ensuring proper gene expression and cellular function. For example, RG108 and MG98 are both small molecules that have been developed as inhibitors of DNMTs. However, there are some differences between these two compounds. RG108 is a non-nucleoside inhibitor of DNMTs that works by binding to the catalytic site of the enzyme and blocking its activity. It has been shown to be effective in reducing DNA methylation levels in various cell types and has been used in several studies to investigate the role of DNA methylation in gene expression and other cellular processes. RG108 has also been evaluated for its potential therapeutic applications, including as a treatment for cancer and other diseases that are associated with abnormal DNA methylation patterns. On the other hand, MG98 is a nucleoside-based inhibitor of DNMTs that acts by incorporating into DNA during replication and inhibiting the activity of DNMTs. It has been shown to be effective in reducing DNA methylation levels in cancer cells and has been evaluated in several clinical trials as a potential treatment for various types of cancer, including non-small cell lung cancer and pancreatic cancer. Both RG108 and MG98, are promising compounds for the development of DNMT inhibitors, yet have different mechanisms of action and hence may be more suitable for different applications or disease indications.*

is hemi-methylated following the replication of the DNA [14, 15]. There are several methylating agents capable of methylating cytosine residues such as DNMT3A and DNMT3B, also known as *de novo* methylation [16]. Dinucleotide CpG is separated all over the human genome and are compiled up in regions rich in CpG in some regions with big repetitive sequences [17]. There are many processes of hyper-methylation that induce suppression of transcriptional factor; the first step is the CpG island which is methylated and absorbs proteins which have inhibitory factors proteins that stop the interaction with the transcription factor and with the sequences of DNA. There is also another mechanism the CpG-methyl protein that basically can recognize the methylated CpG and could suppress the activity of the methylated DNA. Over the years, many scientists have actually exposed the fact that the proliferation of tumors and initiation of human cancer is indeed caused by the silencing of various tumor suppression genes which once methylated, will lead to changing the pathways and then will eventually result in carcinogenesis [18].

### 2.2 RNA methylation

Even in cells with the correct DNA sequencing, the RNA may go through changes that would alter which proteins are produced. Herein, clinically, such changes may lower the protein levels that impact killing the cancer cells or even increase the protein levels that prompt a cancer cell to continue dividing.

The modified version of RNA $N^6$-methyladenosine can also be referred to the residues of adenosine at the position of the N-6. It was originally discovered in the 1970s taking epigenetics and cancer biology by storm [19]. Modifications of the $m^6A$ usually start near the stop codon, 3 UTR, and also, within the long internal exons. Such usually affects all the processing of the RNA, including degradation, transcription, translation as well as splicing. Furthermore, there are studies that suggest that $m^6A$ is dynamic and reversible [20]. Briefly, the formation of the $m^6A$ requires multiple methyl-transferases. Examples include: METTL3, METTL16, and also METTL14 [21].

#### *2.2.1 Is "$m^6A$" the most common RNA modification involved in various cancers?*

The most common modification in RNA is $m^6A$, confirmed worldwide, in various yeasts, drosophila, mammals, viruses and are involved in various aspects of biology and medicine [22]. The modification of the RNA $m^6A$ can lead to different consequences such as affecting mitosis, cell division, gametogenesis, immune homeostasis, and a different biological rhythm [23]. In recent studies, the modification of $m^6A$ is investigated elaborately, primarily in mammals. The methylation of the $m^6A$ RNA takes place within a sequence of purine [G>A] $m^6AC$ [A/U/C], but $m^6A$ is not situated at all methylated sites [24]. Further, the modification of $m^6A$ takes part in different pathogenesis of several other human diseases; particularly in cancer. Indeed, the modification of $m^6A$ RNA is so dynamic that it plays important role in cancer progression and carcinogenesis [25]. As a target, modification of $m^6A$ is taken as a potential tumor marker for diagnosis. There are mainly three components for $m^6A$ regulation: the $m^6A$ methyltransferases ("writers"), $m^6A$ demethylases ("erasers"), and the binding protein decoder called $m^6A$ methylation ("readers"). Also, it is noteworthy that $m^6A$ is present in different forms of RNA but not present in the small nuclear RNA, ribosome RNA, long non-coding RNA, and circular RNA [26]. There are several enzymes involved in $m^6A$ modification: methyl transferase complex, which is a composition of several components including methyltransferase-like protein 14 (METTL14), methyltransferase-like protein 3 (METTL3), and Wilms' tumor 1–associated protein (WTAP) [27]. The earliest known of these three enzymes is the "$m^6A$-writers". Some of the newly discovered writer proteins are zinc finger CCCH domain-containing protein 13 (ZC3H13) [28], methyltransferase-like protein 16 (METTL16) [29], KIAA1429 (VIRMA) [30], CBLL1 (an E3 ubiquitin ligase) [31], and RNA-binding motif protein 15 (RBM15) [32]. These enzymes play role in different pathways [33] and in RNA modification. Different capabilities were revealed in a recent study where distinct genes were silenced (siRNA) like WTAP, METTL3, and METTL14, in cell lines such as HeLa and 293FT (**Figure 2**) [34].

### 2.3 Histone modifications and modifiers

Histones are a family of small, positively charged proteins that are responsible for packaging DNA into chromatin structures in eukaryotic cells. They play a crucial role in the organization and compaction of DNA, which is necessary to fit the long DNA molecules into the small nucleus of a cell. Histones are divided into five main classes: H1, H2A, H2B, H3, and H4, based on their sequence and structure. They contain a high proportion of positively charged amino acids, particularly lysine and arginine, which enable them to interact with the negatively charged DNA backbone

| Nucleoside Analogue | Regulators | Ref |
|---|---|---|
| $m^6A$ | METTL3 | (22) |
| | METTL14 | |
| | METTL16 | |
| | KIAA1429 | |
| | RBM15 | |
| $m^5C$ | NSun2 | |
| | NSun1/3/7 | |
| | NSun4 | |
| | NSun5 | |
| | NSun6 | |
| $m^1A$ | TRMT61B | |
| | TRM61 | |
| | TRMT10C | |
| | TRMT6 | |
| | ALKBH1/3 | |
| $m^3C$ | ALKBH3 | |
| | METTL6 | |
| | METTL8 | |
| $m^7G$ | METTL1-WDR4 complex | |

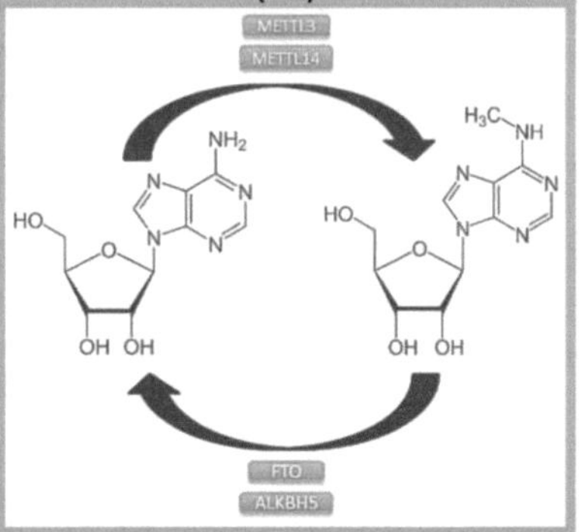


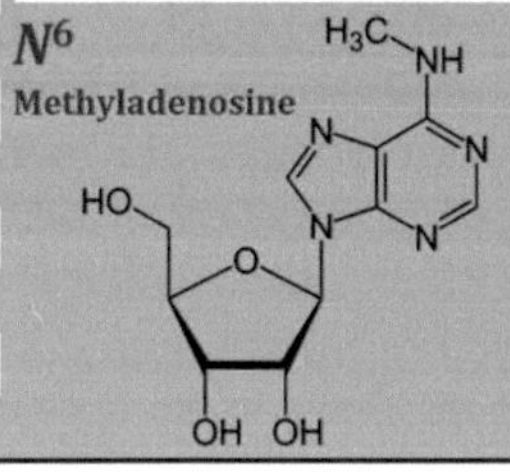


**Figure 2.**
*RNA methylation regulators and $N^6$-methyladenosine ($m^6A$). RNA methylation, particularly N6-methyladenosine ($m^6A$) modification, is an essential post-transcriptional regulatory mechanism that affects RNA stability, splicing, localization, and translation. Several regulators have been identified that control the activity and expression of the $m^6A$ writer and eraser complexes, as well as other factors that affect m6A modification. These regulators can be classified into three main categories: (1) writer complex regulators, (2) eraser complex regulators, and (3) reader protein regulators. The $m^6A$ modification is added to RNA molecules by a complex of proteins known as the $m^6A$ writer complex, which includes methyltransferase-like 3 (METTL3), methyltransferase-like 14 (METTL14), and Wilms tumor 1-associating protein (WTAP). The $m^6A$ modification is removed by a complex of proteins known as the m6A eraser complex, which includes fat mass and obesity-associated protein (FTO) and alpha-ketoglutarate-dependent dioxygenase homolog 5 (ALKBH5). Writer complex regulators include proteins such as RNA-binding motif protein 15/15B (RBM15/15B), which enhances the activity of the writer complex, and heterogeneous nuclear ribonucleoprotein C (HNRNPC), which competes with the writer complex for RNA binding sites and can inhibit $m^6A$ modification. Eraser complex regulators include proteins such as insulin-like growth factor 2 mRNA-binding protein 2 (IGF2BP2), which stabilizes m6A-modified RNA and can protect it from eraser complex-mediated demethylation. Reader protein regulators include proteins such as YTH domain-containing proteins (YTHDFs), which bind to m6A-modified RNA and affect RNA stability and translation. The regulation of RNA methylation is a complex process that involves multiple factors, and a better understanding of these regulators and their interactions is critical for elucidating the functional roles of RNA methylation in various biological processes and diseases. Indeed, recent studies have shown that dysregulation of $m^6A$ modification and its regulators can contribute to various diseases, including cancer, neurological disorders, and viral infections. Therefore, to re-emphasize, understanding the mechanisms of RNA methylation and its regulation by different factors is crucial for developing novel therapeutic approaches targeting this pathway.*

and form tight complexes. Histones can be modified by various post-translational modifications, such as acetylation, methylation, phosphorylation, ubiquitination, and sumoylation, which can affect their structure, function, and interactions with other proteins and DNA. These modifications are critical for regulating gene expression, DNA replication, DNA repair, and other cellular processes. Histone modification is an important post-translational process that plays a key role in gene expression. The modifications impact this gene expression by changing the structure of chromatin or through the recruitment of histone modifiers. The modification of histones, proteins that bind to the DNA, has been documented to be involved in the pathogenesis of autoimmune diseases, including rheumatoid arthritis, systemic lupus erythematosus, systemic sclerosis, primary biliary cirrhosis, and type 1 diabetes. Basically, histones are highly basic proteins abundant in the lysine and arginine residues that are found in the eukaryotic cell nuclei. Histones act as spools around which the DNA winds to create the structural units called nucleosomes, which in turn are wrapped into fibers (30 nm) that form the tightly-packed chromatin. The DNA in the chromatin which is packed in a highly compact structure, and which is wrapped with histones where it forms a nucleosome structure; is also referred to as "beads on a string". This also helps in controlling the ability to access the DNA sequence [35]. Briefly, all histone octamer consists of tetramer two copies of histones 2A and also two copies of histones 2B, bounded by histone 4 and histone 3. There are five types of histones namely H2A, H2B, H3, H4, and H1 linker histone. These proteins generally consist of a C-terminal domain and a tail of a terminal N that results in post-translational modification are acetylation, phosphorylation, methylation, SUMOylation, citrullination, and also biotinylation at some of the specific amino acid group [36]. Among the post-translational modification, acetylation and methylation are the most studied on the residues H3 and H4 [37]. Histone modifications impact gene expression either via changing the structure or regulating the physical properties of chromatin or through the recruitment of histone modifiers (**Figure 3**).

| HYDROXAMATE | ALIPHATIC ACID | BENZAMIDE | CYCLIC TETRAPEPTIDE | Ref |
|---|---|---|---|---|
| SUBEROYLANILIDE HYDROXAMIC ACID (SAHA, VORINOSTAT) | Valproic acid (VPA) | MGCD0103 | Depsipeptide (FK228) | (46) |
| TRICHOSTATIN A | Baceca | MGCD0103 MS-275 (SNDX-275) | CHAPs | (14) |
| LAQ824, LBH589 | Savicol | | Apicidin | (47) |
| PXD101 | AN-9 (prodrug) | | Trapoxin A | |
| OXAMFLATIN, SCRIPTAID, SBHA | Phenylbutyrate | | --- | |
| SK-7041, SK-7068 | --- | | --- | |
| PYROXAMIDE | --- | | --- | |
| TUBACIN | --- | | --- | |

**Figure 3.**
*Examples of histone deacetylase inhibitors. Histone deacetylase inhibitors (HDAC inhibitors) are a class of compounds that interfere with the activity of histone deacetylases, which are enzymes that remove acetyl groups from histone proteins. By inhibiting these enzymes, HDAC inhibitors increase the acetylation of histones and other proteins, leading to changes in chromatin structure and gene expression. HDAC inhibitors have been studied for their potential therapeutic applications in various diseases, including cancer, neurodegenerative disorders, and inflammatory diseases. Some examples of HDAC inhibitors include: (1) Vorinostat (also known as SAHA): This was the first HDAC inhibitor to be approved by the US FDA for the treatment of cutaneous T-cell lymphoma. It works by inhibiting class I and II HDACs. (2) Romidepsin: This is another HDAC inhibitor that is approved for the treatment of cutaneous T-cell lymphoma. It works by inhibiting class I HDACs. (3) Belinostat: This is an HDAC inhibitor that is approved for the treatment of peripheral T-cell lymphoma. It works by inhibiting class I, II, and IV HDACs. (4) Panobinostat: This is an HDAC inhibitor that is approved for the treatment of multiple myeloma. It works by inhibiting class I, II, and IV HDACs. (5) Entinostat: This is an HDAC inhibitor that is being investigated for its potential use in breast cancer, lung cancer, and other types of cancer. It works by inhibiting class I HDACs. (6) Trichostatin A: This is a natural product that was the first HDAC inhibitor to be discovered. It works by inhibiting class I and II HDACs. These are just a few examples of the many HDAC inhibitors that have been developed and studied over the years. Each inhibitor has its unique characteristics, including its peculiar mechanism of action, target specificity, and potential therapeutic applications.*

## 3. Role of histone modification and associated inhibitors in tumorigenesis

Cancer is a complex disease, and the first human disease to be correlated with epigenetic alterations. As mentioned throughout this chapter, it is today clear that epigenetics plays a key role in tumorigenesis (tumor development and metastasis) via the regulation (and control) of gene expression. This is mainly through DNA methylation, histone modifications (established and removed by the modifier enzymes: *writers* and *erasers*, respectively), histone variant incorporation, chromatin remodeling, and non-coding RNAs. Briefly, mutations within the chromatin remodeling complexes or the histones affect the cell phenotype, thereby, leading to various human diseases, including cancer. Indeed, with the promoter-targeted histone modification, the studies have recently shown that the modification of histones at the cellular level is highly related to the prognosis of cancer. The initial interest evolved when there was a histone modification derived from yeast [38], and all the patterns of the histone modification at the promoter to the expression of the gene in the yeast suggested that the histone can keep multiple biological information in the pattern of modification [39]. Histone modification takes place throughout the genome, and any possible changes if occur in the specificity or activity of the enzymes (which helps in the modification of the histones) can result in changes. Those are detectable at many specific modifications of the histones and at the level of the individual nuclei, in the process of immunostaining [40].

It is perhaps worth mentioning herein, that while there are several approved (by the US Food and Drug Administration or FDA) epigenetic-based drugs or *epi-drugs* to treat cancer with abnormal histone modifications, many are still in the pre- or clinical phase. Such indicates the need to better understand the regulatory pattern of epigenetics, especially histone modifications, in cancer. The modification of histones suggests replacing "histone code" to follow and maintain the interaction of histones with a protein associated with chromatin and allow all the downstream functions [41]. Consistently, the HATs enzymes which are known as the "writers" transfer the acetyl groups to some targeted lysine group and arginine group residues in the histone tails, which often results in the activation of the gene [42]. Also, the well-known "erasers" and the HDAC enzyme often delete the acetyl group from the histone tail and downregulate the targeted gene. The "writers" and "erasers" can modify the histones and control the silent and active state of chromatin and thus can transcriptionally control the transcription of all genetic information present in the DNA [43]. For treatment and clinical diagnosis, therapeutic targets are key [44]. Various modified drugs are useful to block cancer metastasis and the progression of the tumor via a major impact on the modification of histones, mainly via histone acetylation [45, 46]. Indeed, the dysregulation of histone modification enzymes plays an important role in tumorigenesis. Herein, miRNA acts directly on two pathways, the translational silencing mechanism, and the transcriptional silencing mechanism [47, 48]. In one of these cases, the miRNA ties up with the sequence, for example, mutation, and the regulation of the main gene gets out of control and can often result in chemo-resistance and tumorigenesis, subsequently [49]. For example, in ovarian cancer, the regulation of miRNA is downregulated and the let-7i resistance toward cisplatin is consequently increased with slower progression. The survival time for the patient decreases in the late stage of ovarian cancer [50] and the possible treatment for those patients is gene therapy; the only way to silence aberrant miRNA expression or restore the lost endogenous miRNA expression. Therefore, therapeutically, this can often be accompanied by an approved epi-drug [51]. Remember, both, acetylation or

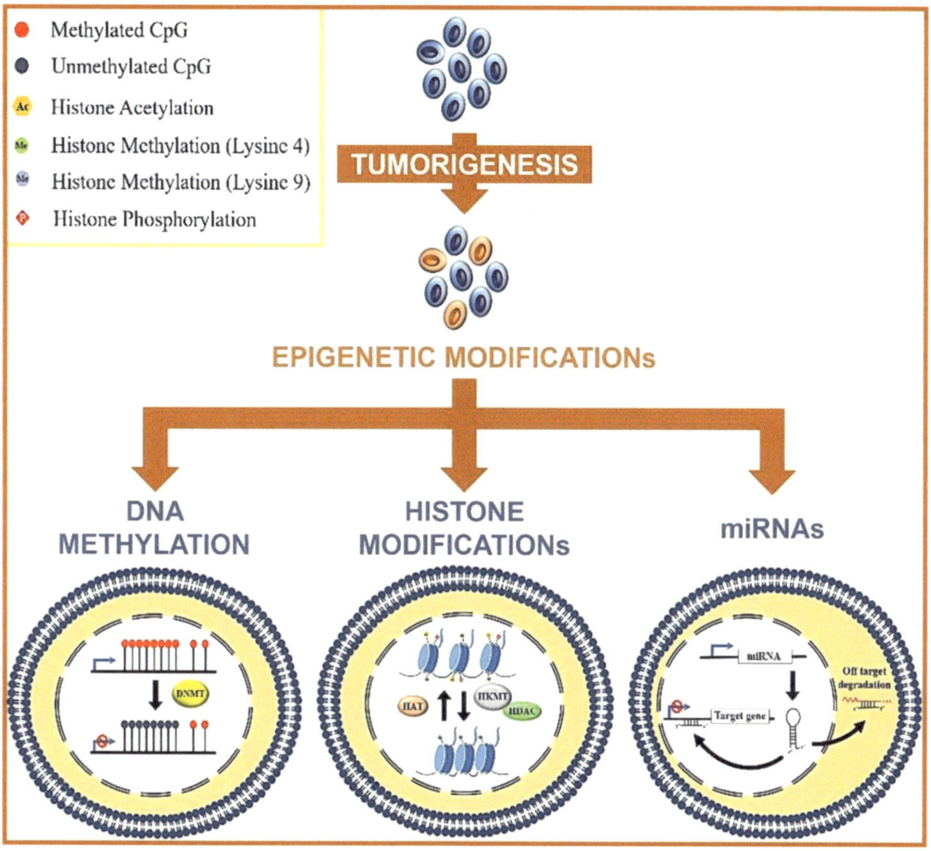


**Figure 4.**
*Epigenetic regulation in tumorigenesis. Epigenetic regulation plays a critical role in tumorigenesis, which is the process of tumor formation. Alterations in DNA methylation, histone modifications, and non-coding RNA expression can lead to abnormal gene expression patterns that contribute to tumor development and progression. Epigenetic changes can affect various cellular processes, including cell proliferation, apoptosis, DNA repair, and immune response, among others. Additionally, epigenetic alterations are reversible, making them an attractive target for developing new cancer therapies. Hence, understanding the epigenetic changes associated with tumorigenesis is deemed crucial for identifying new targets for cancer treatment and developing personalized therapies based on the unique epigenetic profile of the patient.*

deacetylation of histone proteins regulates gene expression, and the combination of epi-drugs together or with other inhibitors has displayed favorable clinical outcomes (**Figure 4**).

Histone modification is a key step in gene regulation that determines cell fate. For epigenetic therapy, the main target is chemo-resistant cells. DNA methylation that occurs at GHD CpG islands can often result in the inactivation of the transcriptional gene which is highly present in tumors. Inhibition of some enzymes such as DNMTs (which catalyze the methylation of the DNA), results in a decrease of the DNA methylating agent, and therefore re-activates the genes which are potentially anti-cancerous [52]. Modification of histones is un-altered via some special set of enzymes including the HDACs and HKMTs. Furthermore, targeting atypical hypo-acetylation through HDACi can result in and/or lead to the re-activation of the foregoing and transcriptionally inadequate chromatin [53]. Likewise, the suppression of the special atypical histone methyltransferases or HKMTs stops the methyl marks that may cause gene repression; HP1 to H3K9me3, for example. Two vital processes of epigenetics are: histone modification and DNA methylation. Also, remember that atypical expression of miRNA (*21–23 nucleotides small and non-protein-coding RNAs*) has a direct connection with tumorigenesis [54]. Cancer is a complex and systemic disease instead of a single organ or tissue failure. Therefore, a single drug cannot solely treat or cure the tumor completely; often resulting in tumor recurrence or resistance. Precision oncology medicine is the hope.

## 4. Mechanisms of acquired anti-tumor and -cancer drug resistance

Resistance to anti-cancer drugs can be acquired by several mechanisms within neoplastic cells. Some include the alteration of drug targets, expression of drug pumps, expression of de-toxification mechanisms, reduced susceptibility to apoptosis, and increased ability to repair DNA damage, among others. Similarly, anti-tumor drug resistance of cancer cells is chiefly acquired through one of the three mechanisms of (a) mutation in gene, (b) gene expression increase, and (c) decrease in gene expression [55]. An example of anti-tumor drugs is the PI (Protease Inhibitor)-based drugs which act by inhibiting important cell signaling pathways [56]. Mono-therapy with PI has failed miserably, nonetheless, in patients with multiple myeloma (MM), for instance, due to the development of slow drug tolerance. It is also a common observation that many ovarian cancer patients treated with only platinum-based drugs do change into refractory cancer from being advanced and recurrent [57]. It has also been reported that cisplatin resistance is associated with hypomethylation of CpG sites in the first intron of S100A4 [58]. Anti-tumor drug resistance is associated with down-regulation of tumor protein p53 (TP53) and interferon regulatory factor 1 (IRF1) and activation of HDAC [59]. Hence, to avoid the development of drug tolerance in MM patients combination therapy (HDAC inhibitors + PI-based therapy) has shown an overall good therapeutic effect and negated the drug tolerance pathways [60]. Remember that resistance can occur when even a small group of cancer cells within a tumor contain or undergo molecular changes rendering them insensitive to a or any specific drug before the oncology treatment even begins (**Figure** 5).

## 5. Targeting of epigenetic modifiers via acting against drug resistance

Today, despite the availability of several potential drugs targeting HDACs (histone deacetylases) and DNMT/HMT (DNA/histone methyltransferases) for treating a wide range and types of cancers; yet, are often limited in efficacy to function only at certain stages of the disease. Indeed, in some cases and even after the drug treatment, the reversible nature of methylation persists. Hence why, recent studies tend to increasingly emphasize the need to develop novel drugs capable of specifically targeting the HDACs and DNMT inhibitors as an emerging bio-effective anti-cancer strategy. On the other hand, there are several studies that have proved that mutation in a gene, and modification in the epigenetics do play a very vital role in chemo-resistance in the cells which are infested by cancer [61]. The alterations in the epigenetic are very much reversible and must be maintained by epigenetic modifiers rendering them a good target for therapeutic interference [62, 63]. The recent research and understanding of the CSC epigenome provide a new perception of anti-cancer therapy which is much targeted by many epigenetic drugs to overcome

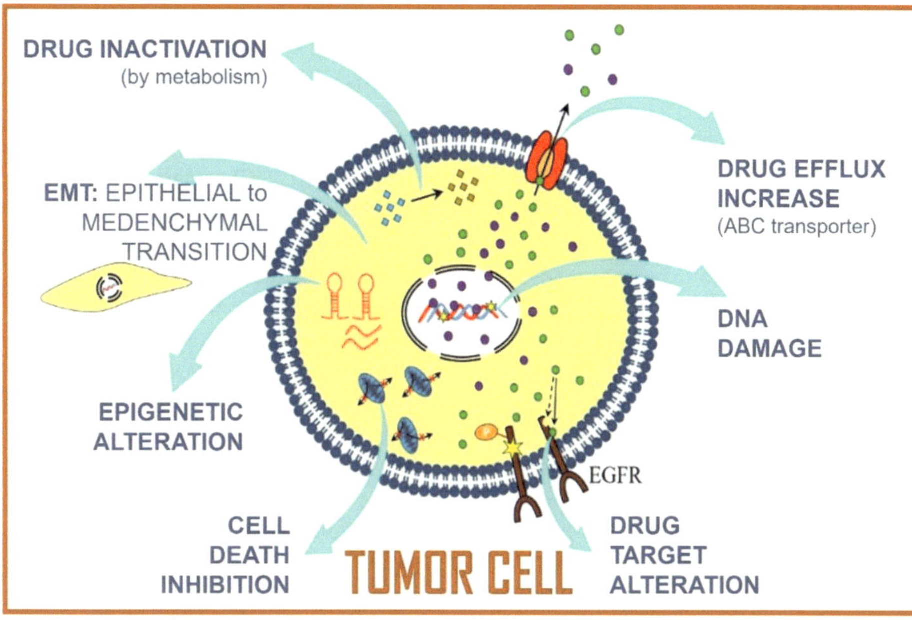


**Figure 5.**
*Epigenetic modifications can be tumorigenic and contribute to the development and progression of cancer. The study of epigenetic modifications in cancer has revealed new insights into the molecular mechanisms of tumorigenesis, as well as potential biomarkers for cancer diagnosis, prognosis, and treatment. Epigenetic changes can affect key pathways involved in DNA damage response, cell cycle regulation, and immune surveillance, leading to genomic instability and immune evasion. Furthermore, epigenetic changes can drive cancer cell heterogeneity, making tumors more resistant to therapy. The reversibility of epigenetic changes makes them an attractive target for developing new and innovative translational cancer therapies. However, there are still significant challenges to developing effective epigenetic therapies, including identifying specific epigenetic targets and minimizing off-target effects. Understanding the complex interplay between epigenetic modifications and cancer is deemed crucial for developing new strategies to prevent, diagnose, and treat cancer. Indeed, it is worth noting that epigenetic modifications are not always tumorigenic and can play essential roles in normal cellular processes, including development and differentiation. Remember that epigenetic regulation is a complex and dynamic process that involves the interplay of various enzymes, chromatin-associated proteins, and non-coding RNAs. The dysregulation of these epigenetic regulators can lead to aberrant gene expression patterns that contribute to tumorigenesis. Therefore, identifying the specific epigenetic modifications that contribute to cancer development and progression is henceforth deemed critical for developing novel targeted therapies that can "selectively" reverse these changes and later, restore normal cellular function.*

resistance toward CSC drugs [64], as mentioned earlier. Accruing research tends to show that drugs with epigenome modifying capabilities or epi-drugs, alone or combined with other treatments may/can modify the epigenetic treatment and decrease the resistance toward the drug. Also, drugs which have the capability to alter the genome are "non-specific" in nature and can affect the expression of the global-gene, with increasing evidence in the literature that such drugs can make changes in gene expression depending on the chromatin environment [65]. Furthermore, other studies demonstrated that the sensitivity toward epigenetic alternators can be genomic loci specific depending upon the three-dimensional structure of the chromatin itself [66]. As noted earlier, there are various epigenetic modifiers which are mostly used in the present-time anti-cancer therapy clinical trials and research, i.e., HDACs, DNMTs, and HMTs [67]. Noteworthy that various inhibitors used for HDACs, DNMTs, histone demethylases (HDMs), HMTs and bromodomain proteins are generally, in clinical trial studies, used in combination with or without chemotherapy [68].

## 6. Conclusions

Cancer is a complex disease. For carcinogenesis and tumorigenesis, epigenetic changes are fundamental mechanisms and can serve as potential methods for early detection, treatment, and prognostic assessment for our oncology patients. Epigenetics was first introduced by Conrad Waddington in 1942 to define stable and heritable changes in the cell phenotype and gene expression without genetic alterations or DNA sequence. Today, epigenetic modifications and processes include DNA and RNA methylation, histone covalent modifications, chromatin remodeling, and the effect of non-coding RNAs and polycomb proteins in gene expression. Herein, and in the landscape of epigenetics, the methylation of DNA and RNA and modifications of histone in the cancerous cells may be responsible for drug resistance and the recurrence of cancer. While epigenetic changes may be used as tools to diagnose, treat, and provide prognostic information for our cancer patients, a better understanding of such mechanisms associated with epigenetic modifications; an ongoing investigation and research effort, will eventually not only result in the development of new epigenetic bio-markers capable of early detection of tumors and the maintenance of continuous surveillance but also in the identification of distinct epigenetic profiles that will corroborate to the identification of prognostic tools and a potential predictor of tumor response to therapy. Further, as we move toward personalized and precise medicine, it is important to remember that the genetic material is identical in every cell, while epigenetics is highly variable within different cells and tissues of an organism and is also affected by aging and environmental factors. Herein, identifying and validating such novel epigenetic modifications associated with cancer chemoresistance, in clinical studies, cannot be underestimated. Together, the discovery of new epigenetic biomarkers for individual cancer chemo-resistance can and will open the possibility for the development of novel epi-drugs, which can be used as an adjuvant therapy associated with conventional chemotherapeutic drugs, enhancing tumor sensitivity to traditional agents and ultimately increasing therapeutic efficacy. Regardless of the type or location of the cancer, epigenetic modifications induced by epi-drugs, aligned with the prospective identification of epigenetic biomarkers, are an exciting frontier in cancer biology, awaiting to be explored.

## Acknowledgements

This work was supported by operating grants provided to the HAiDAR R&D&I LAB/BioMAT'X (Laboratorio de Biomateriales, Farmacéuticos y Bioingeniería de Tejidos Cráneo Máxilo-Facial), member of CiiB (Centro de Investigación e Innovación Biomédica), Faculties of Medicine and Dentistry, Universidad de los Andes, Santiago de Chile, through the ANID-NAM (Agencia Nacional de Investigación y Desarrollo, Chile and National Academy of Medicine, USA) Grant código # NAM21I0022 (2020–2022), CORFO Crea y Valida I+D+i Grant código # 21CVC2-183649 (2021–2023), CORFO Crea y Valida-Proyecto de I+D+i Colaborativo-Reactívate" Grant código # 22CVC2-218196 (2022–2024), and FONDEF Concurso IDEA de I+D, ANID, Grant código # ID22I10215 (2022–2024). The author wishes to acknowledge the exceptional UAndes F-ODO pre-graduate students behind stirring this piece: Yr 2 (Camila Alhucema, Valeria González and Benjamin Mora), Yr 4 (Andrea Bustos, Ismael Valenzuela and Zabdiel Faundez), Yr 5 (Alondra Beniscelli), and Yr 6 (Ignacio Fernández).

## Conflict of interest

The author declares no conflict of interest.

## Notes/thanks/other declarations

I would like to thank Ms. Dolores Kuzelj, Author Service Manager at IntechOpen for her constant support, communication, feedback, and in the timely preparation of our present book project "DNA Replication—Mechanisms, Epigenetics, and Gene Therapy Applications", with content, to the best of abilities, *different* from what has been previously published in the literature. I THANK YOU and hope for the next book.

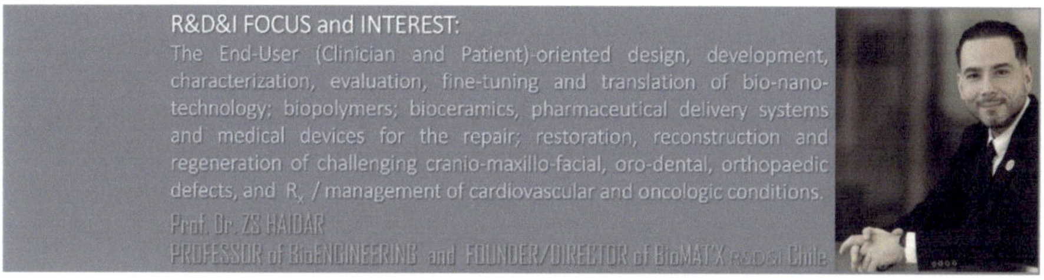


## Author details

Ziyad S. Haidar
BioMAT'X (HAiDAR R&D&I LAB) I+D+i Labratory, Biomedical Research and Innovation Center/Centro de Investigación e Innovación Biomédica (CiiB), Faculties of Medicine and Dentistry, Universidad de los Andes, Las Condes, Santiago, Chile

*Address all correspondence to: zhaidar@uandes.cl; zhaidar78@gmail.com

## References

[1] Aziz MH, Ahmad A. Epigenetic Basis of Cancer Drug Resistance. CDR [Internet]. 2020. [cited 2022 Oct 15]. Available from: https://cdrjournal.com/article/view/3374

[2] Hayashi T, Konishi I. Correlation of anti-tumour drug resistance with epigenetic regulation. British Journal of Cancer. 2021;**124**(4):681-682

[3] Chen HT, Liu H, Mao MJ, Tan Y, Mo XQ, Meng XJ, et al. Crosstalk between autophagy and epithelial-mesenchymal transition and its application in cancer therapy. Molecular Cancer. 2019;**18**(1):101

[4] Sánchez-Tilló E, Liu Y, de Barrios O, Siles L, Fanlo L, Cuatrecasas M, et al. EMT-activating transcription factors in cancer: Beyond EMT and tumor invasiveness. Cellular and Molecular Life Sciences. 2012;**69**(20):3429-3456

[5] Ramesh V, Brabletz T, Ceppi P. Targeting EMT in cancer with repurposed metabolic inhibitors. Trends Cancer. 2020;**6**(11):942-950

[6] Wei J, Lu Y, Wang R, Xu X, Liu Q, He S, et al. MicroRNA-375: Potential cancer suppressor and therapeutic drug. Bioscience Reports. 2021;**41**(9):BSR20211494

[7] Liu J, Yan W, Han P, Tian D. The emerging role of KIAA1199 in cancer development and therapy. Biomedicine & Pharmacotherapy. 2021;**138**:111507

[8] Wilson MM, Weinberg RA, Lees JA, Guen VJ. Emerging mechanisms by which EMT programs control stemness. Trends Cancer. 2020;**6**(9):775-780

[9] Dawson MA, Kouzarides T. Cancer epigenetics: From mechanism to therapy. Cell. 2012;**150**(1):12-27

[10] Lu Y, Chan YT, Tan HY, Li S, Wang N, Feng Y. Epigenetic regulation in human cancer: The potential role of epi-drug in cancer therapy. Molecular Cancer. 2020;**19**(1):79

[11] Kulis M, Esteller M. DNA methylation and cancer. Advances in Genetics. 2010;**70**:27-56

[12] Zafon C, Gil J, Pérez-González B, Jordà M. DNA methylation in thyroid cancer. Endocrine-Related Cancer. 2019;**26**(7):R415-R439

[13] Nishiyama A, Nakanishi M. Navigating the DNA methylation landscape of cancer. Trends in Genetics. 2021;**37**(11):1012-1027

[14] Zeller C, Brown R. Therapeutic modulation of epigenetic drivers of drug resistance in ovarian cancer. Therapeutic Advances in Medical Oncology. 2010;**2**(5):319-329

[15] Jones PA, Baylin SB. The fundamental role of epigenetic events in cancer. Nature Reviews. Genetics. 2002;**3**(6):415-428

[16] Jia D, Jurkowska RZ, Zhang X, Jeltsch A, Cheng X. Structure of Dnmt3a bound to Dnmt3L suggests a model for de novo DNA methylation. Nature. 2007;**449**(7159):248-251

[17] Rodríguez-Paredes M, Esteller M. Cancer epigenetics reaches mainstream oncology. Nature Medicine. 2011;**17**(3):330-339

[18] Pan Y, Liu G, Zhou F, Su B, Li Y. DNA methylation profiles in cancer diagnosis and therapeutics. Clinical and Experimental Medicine. 2018;**18**(1):1-14

[19] An Y, Duan H. The role of m6A RNA methylation in cancer metabolism. Molecular Cancer. 2022;**21**(1):14

[20] Pan Y, Ma P, Liu Y, Li W, Shu Y. Multiple functions of m6A RNA methylation in cancer. Journal of Hematology & Oncology. 2018;**11**(1):48

[21] Lan Q, Liu PY, Bell JL, Wang JY, Hüttelmaier S, Zhang XD, et al. The emerging roles of RNA m6A methylation and demethylation as critical regulators of tumorigenesis, drug sensitivity, and resistance. Cancer Research. 2021;**81**(13):3431-3440

[22] Jonkhout N, Tran J, Smith MA, Schonrock N, Mattick JS, Novoa EM. The RNA modification landscape in human disease. RNA. 2017;**23**(12):1754-1769

[23] Yang Y, Hsu PJ, Chen YS, Yang YG. Dynamic transcriptomic m6A decoration: Writers, erasers, readers and functions in RNA metabolism. Cell Research. 2018 Jun;**28**(6):616-624

[24] Dominissini D, Moshitch-Moshkovitz S, Schwartz S, Salmon-Divon M, Ungar L, Osenberg S, et al. Topology of the human and mouse m6A RNA methylomes revealed by m6A-seq. Nature. 2012;**485**(7397):201-206

[25] Sharma S, Kelly TK, Jones PA. Epigenetics in cancer. Carcinogenesis. 2010;**31**(1):27-36

[26] Huang H, Weng H, Chen J. The Biogenesis and Precise Control of RNA m6A Methylation. Trends in Genetics. 2020;**36**(1):44-52

[27] Bokar JA, Shambaugh ME, Polayes D, Matera AG, Rottman FM. Purification and cDNA cloning of the AdoMet-binding subunit of the human mRNA (N6-adenosine)-methyltransferase. RNA. 1997;**3**(11):1233-1247

[28] Wen J, Lv R, Ma H, Shen H, He C, Wang J, et al. Zc3h13 regulates nuclear RNA m6A methylation and mouse embryonic stem cell self-renewal. Molecular Cell. 2018;**69**(6):1028-1038.e6

[29] Mendel M, Chen KM, Homolka D, Gos P, Pandey RR, McCarthy AA, et al. Methylation of structured RNA by the m6A writer METTL16 is essential for mouse embryonic development. Molecular Cell. 2018;**71**(6):986-1000.e11

[30] Cheng X, Li M, Rao X, Zhang W, Li X, Wang L, et al. KIAA1429 regulates the migration and invasion of hepatocellular carcinoma by altering m6A modification of ID2 mRNA. OncoTargets and Therapy. 2019;**12**:3421-3428

[31] Liu T, Li C, Jin L, Li C, Wang L. The Prognostic Value of m6A RNA Methylation Regulators in Colon Adenocarcinoma. Medical Science Monitor. 2019;**25**:9435-9445

[32] Knuckles P, Lence T, Haussmann IU, Jacob D, Kreim N, Carl SH, et al. Zc3h13/Flacc is required for adenosine methylation by bridging the mRNA-binding factor Rbm15/Spenito to the m 6 A machinery component Wtap/Fl(2)d. Genes & Development. 2018;**32**(5-6):415-429

[33] Esteller M. Epigenetics in cancer. The New England Journal of Medicine. 2008;**358**(11):1148-1159

[34] Liu J, Yue Y, Han D, Wang X, Fu Y, Zhang L, et al. A METTL3–METTL14 complex mediates mammalian nuclear RNA N6-adenosine methylation. Nature Chemical Biology. 2014;**10**(2):93-95

[35] Audia JE, Campbell RM. Histone modifications and cancer. Cold Spring Harbor Perspectives in Biology. 2016;**8**(4):a019521

[36] Wang R, Xin M, Li Y, Zhang P, Zhang M. The functions of histone modification enzymes in cancer. Current Protein & Peptide Science. 2016;**17**(5):438-445

[37] Sun T, Liu Z, Yang Q. The role of ubiquitination and deubiquitination in cancer metabolism. Molecular Cancer. 2020;**19**(1):146

[38] Seligson DB, Horvath S, Shi T, Yu H, Tze S, Grunstein M, et al. Global histone modification patterns predict risk of prostate cancer recurrence. Nature. 2005;**435**(7046):1262-1266

[39] Kurdistani SK, Tavazoie S, Grunstein M. Mapping global histone acetylation patterns to gene expression. Cell. 2004;**117**(6):721-733

[40] Kurdistani SK. Histone modifications as markers of cancer prognosis: A cellular view. British Journal of Cancer. 2007;**97**(1):1-5

[41] Lund AH, van Lohuizen M. Epigenetics and cancer. Genes & Development. 2004;**18**(19):2315-2335

[42] Kristensen LS, Nielsen HM, Hansen LL. Epigenetics and cancer treatment. European Journal of Pharmacology. 2009;**625**(1-3):131-142

[43] Feng J, Meng X. Histone modification and histone modification-targeted anti-cancer drugs in breast cancer: Fundamentals and beyond. Frontiers in Pharmacology. 2022;**13**:946811

[44] Elsheikh SE, Green AR, Rakha EA, Powe DG, Ahmed RA, Collins HM, et al. Global histone modifications in breast cancer correlate with tumor phenotypes, prognostic factors, and patient outcome. Cancer Research. 2009;**69**(9):3802-3809

[45] Zhuang J, Huo Q, Yang F, Xie N. Perspectives on the role of histone modification in breast cancer progression and the advanced technological tools to study epigenetic determinants of metastasis. Frontiers in Genetics. 2020;**11**:603552

[46] Fan W, Zhang L, Jiang Q, Song W, Yan F, Zhang L. Histone deacetylase inhibitor based prodrugs. European Journal of Medicinal Chemistry. 2020;**203**:112628

[47] Matei DE, Nephew KP. Epigenetic therapies for chemoresensitization of epithelial ovarian cancer. Gynecologic Oncology. 2010;**116**(2):195-201

[48] Kanwal R, Gupta S. Epigenetic modifications in cancer. Clinical Genetics. 2012;**81**(4):303-311

[49] Nirmaladevi R. Epigenetic alterations in cancer. Frontiers in Bioscience. 2020;**25**(6):1058-1109

[50] Shukla S, Meeran SM. Epigenetics of cancer stem cells: Pathways and therapeutics. Biochimica et Biophysica Acta. 2014;**1840**(12):3494-3502

[51] Humphries B, Wang Z, Yang C. MicroRNA regulation of epigenetic modifiers in breast cancer. Cancers (Basel). 2019;**11**(7):897

[52] Kinnaird A, Zhao S, Wellen KE, Michelakis ED. Metabolic control of epigenetics in cancer. Nature Reviews. Cancer. 2016;**16**(11):694-707

[53] Verma M, Srivastava S. Epigenetics in cancer: Implications for early detection and prevention. The Lancet Oncology. 2002;**3**(12):755-763

[54] Herceg Z. Epigenetics and cancer: Towards an evaluation of the impact

of environmental and dietary factors. Mutagenesis. 2007;**22**(2):91-103

[55] Ellis L, Atadja PW, Johnstone RW. Epigenetics in cancer: Targeting chromatin modifications. Molecular Cancer Therapeutics. 2009;**8**(6): 1409-1420

[56] Stahl M, Kohrman N, Gore SD, Kim TK, Zeidan AM, Prebet T. Epigenetics in cancer: A hematological perspective. PLoS Genetics. 2016;**12**(10): e1006193

[57] Housman G, Byler S, Heerboth S, Lapinska K, Longacre M, Snyder N, et al. Drug resistance in cancer: An overview. Cancers. 2014;**6**(3):1769-1792

[58] Roberti A, Valdes AF, Torrecillas R, Fraga MF, Fernandez AF. Epigenetics in cancer therapy and nanomedicine. Clinical Epigenetics. 2019;**11**(1):81

[59] Villanueva L, Álvarez-Errico D, Esteller M. The contribution of epigenetics to cancer immunotherapy. Trends in Immunology. 2020;**41**(8): 676-691

[60] Rivlin N, Brosh R, Oren M, Rotter V. Mutations in the p53 tumor suppressor gene: Important milestones at the various steps of tumorigenesis. Genes & Cancer. 2011;**2**(4):466-474

[61] Park JW, Han JW. Targeting epigenetics for cancer therapy. Archives of Pharmacal Research. 2019;**42**(2):159-170

[62] Chik F, Szyf M, Rabbani SA. Role of epigenetics in cancer initiation and progression. Advances in Experimental Medicine and Biology. 2011;**720**:91-104

[63] Camuzi D, de Amorim ÍSS, Ribeiro Pinto LF, Oliveira Trivilin L, Mencalha AL, Soares Lima SC. Regulation is in the air: The relationship between hypoxia and epigenetics in cancer. Cell. 2019;**8**(4):300

[64] Keyvani-Ghamsari S, Khorsandi K, Rasul A, Zaman MK. Current understanding of epigenetics mechanism as a novel target in reducing cancer stem cells resistance. Clinical Epigenetics. 2021;**13**(1):120

[65] Ahuja N, Sharma AR, Baylin SB. Epigenetic therapeutics: A new weapon in the war against cancer. Annual Review of Medicine. 2016;**67**(1):73-89

[66] Zhang T, Pilko A, Wollman R. Loci specific epigenetic drug sensitivity. Nucleic Acids Research. 2020;**48**(9):4797-4810

[67] Johnson C, Warmoes MO, Shen X, Locasale JW. Epigenetics and cancer metabolism. Cancer Letters. 2015;**356** (2 Pt A):309-314

[68] Adhikari S, Bhattacharya A, Adhikary S, Singh V, Gadad SS, Roy S, et al. The paradigm of drug resistance in cancer: An epigenetic perspective. Bioscience Reports. 2022;**42**(4): BSR20211812

Chapter 3

# Perspective Chapter: Epigenetic Therapy – The Future Treatment for Cancer

*Surya Chandra Rao Thumu, Shobha Rani Papanna and Sundru Manjulata Devi*

## Abstract

Scientists have made a remarkable breakthrough by uncovering DNA and its role in living organisms. Epigenetics examines the phenotypic divergences due to DNA methylation and its effects at certain genetic spots. Epigenetic and genetic problems combine to cause cancer and its growth, as seen by frequent mutations in genes that manage the epigenome. Recently, new therapies targeting epigenetic alterations have been proposed. Drugs with longer shelf life and better absorption are also being manufactured and tested. On this aspect, CRISPR technology has been used to create various strategies for epigenetic engineering and is a practical approach to understanding and manipulating biological processes. Furthermore, studies on the advantages of probiotics have advanced previous interventional studies to recognize the molecular mechanisms involved. Numerous probiotic genomes include epigenetic components that influence gene expression for fundamental functions. Consequently, we suggest investigations incorporating genomic and meta-epigenomic information to better understand the mode of action of probiotics and their related microbiomes in epigenetic therapy. Here, we review established epigenetic discoveries, combined with the rapid advancement of immunotherapies, to create new possibilities for cancer treatment.

**Keywords:** epigenetics, epigenomes, CRISPR-cas, probiotics, microbiome, DNA methylation

## 1. Introduction

One of the greatest discoveries made in the twentieth century is recognizing that phenotypic changes are not merely based on genetic alterations but are implications due to external stimuli/epigenetic factors [1]. Epigenetic trait is a phenotype whose inheritance is stably heritable without altering the genetic code. Thus, for the normal development of an organism, regulations in the epigenome or epigenetic state are equally important as changes in gene expression. Today, epigenetics is a dynamic field aimed at studying individual and fundamental processes in mitosis and meiosis [2]. Several endogenous factors are found to be involved in modulating epigenetic states and their activity can be affected by each other. Cellular functions can be preserved for an extended period by generating and transmitting epigenetic states [2]. Genomic characteristics such as chromatin reconstitution and epigenetic factors' alterations could play a role in epigenetic states. However, epigenetic changes must be precisely defined to obtain a complete and detailed picture [3–5]. Understanding the role of epigenetic alterations and their effect on gene expression and the awareness that epigenetic traits are reversible has opened the door to developing diagnostic tools to treat several diseases, including cancers [6].

Epigenetic alterations are essential for the regulatory processes associated with several biological phenomena, including genomic imprinting, X-chromosome inactivation, control of tissue-specific gene expression, genomic stability, repression of transposable elements, aging, and some diseases, including cancer [7]. The most common epigenetic modifications that modify DNA accessibility to transcriptional machinery and impact gene expression are acetylation, methylation, phosphorylation, biotinylation, and RNA interference [8]. Roberti et al. [9] have identified three main components of the epigenetic system: DNA methylation, post-translational histone modifications, and noncoding RNAs (ncRNAs). The clustered regularly interspaced short palindromic repeats (CRISPR)-associated protein 9 (Cas9)system (CRISPR-cas9) is a naturally occurring defense system in many bacteria and archaea to protect from invading viruses or plasmids [10]. CRISPR system with the capacity to program Cas9 has revolutionized medical research, biotechnology and agriculture. It is a precise genome editing tool that can modify specific genome regions in eukaryotic cells, particularly mammalian cells [11]. Cas proteins are endonucleases that sever the target nucleic acid sequence. Mutations in genes that encode Cas proteins can generate deactivated cas (dCas) proteins that maintain their ability to bind DNA in a manner targeted by gRNA and serve as programmable DNA binding domains [12]. CRISPRi/a domain proteins can be used in epigenetic studies to alter histone acetylation and methylation [13]. The dCas-based demethylation approach was discovered to improve cell disease conditions. Scientists have found that countering epigenetic changes can be an effective cancer therapy [7, 14].

The trillions of intestinal microflora have an intimate relationship with the host and impact the host's physiology and pathology. They play a role in food fermentation, vitamin synthesis, and maintaining intestinal epithelial function. The microbiome affects the host's epigenetics, immune system, and metabolism. Certain epigenetic changes after exposure to bacteria suggest a connection between the microbiome and the epigenome [15]. Investigating the epigenetics and para-epigenetics of a probiotic organism and its microbiome is a difficult but critical step in understanding the probiotic mechanism [1]. Researchers have used CRISPR-Cas to precisely alter probiotic organisms, explore epigenetics, and improve the effects of probiotics [16, 17]. Certain epigenetic changes during childhood are essential for adult health, and short-chain fatty acids (SCFAs), such as butyrate, can influence human genes by being epigenetic regulators [18, 19]. Researching the various probiotic molecules and their interactions with human epigenetic systems could reveal new health benefits and assess the safety of consuming these organisms. Current and future technologies can help researchers understand microbial meta-epigenomics and how it can improve human health. In the current study, a search strategy was assigned, and the search words with "EPIGENETICS", CRISPR-cas9 EDITING IN EPIGENETIS" and PROBIOTICS THERAPY IN EPIGENETICS" has been used and the related articles were retrieved from PubMed, Web of Science, and Scopus databases from 2010 to Jan 2023. This article discusses epigenetic therapy and its potential new methodologies using the

CRIPSR-cas system, epigenetic drugs, and epigenetic regulation by probiotic bacteria and the gut microbiota.

## 2. The emergence of the CRISPR Cas9 system in epigenome editing

Epigenetic changes at the target site can be achieved by securing the epigenetic effectors to sequence-specific DNA-binding domains (DBDs). Although the discovery of zinc fingers (ZFNs) and TALENs succeeded in DNA binding and genome editing efficacy, targeting different regions required redesigning and reengineering of these proteins. This has prevented the scientific community from using ZFN and TALEN effectively [20]. The rise of the CRISPR Cas system has trounced this major hurdle. Although the Cas9 protein of the type II *Streptococcus pyogenes* system is the most widely used in the CRISPR Cas system in gene knockout, the CRISPR-Cas system in its basic form has become a powerful technology to flexibly target any specific site in the genome simply by employing small guide RNAs (sgRNA). Because most Cas proteins are nucleases, mutations in Cas-encoding genes convert them to deactivated Cas (dCas) proteins, inhibiting their nuclease activity while retaining DNA binding capability. Thus, dCas proteins have become programmable DBP, and the ability to bind dCas9 to specific DNA loci can be altered using sgRNA (**Figure 1**) [12].

### 2.1 CRISPR Cas9 in epigenome activation and repression

Soon after discovering that wild-type Cas9 could be used as a programmable endonuclease, researchers began to exploit dCas9 for gene regulation studies. For epigenome editing through CRISPR Cas9, the setup consists of a toolbox for expressing sgRNA. In these dCas9 and effector domains, the activation domains retain their function when fused to other proteins with different DNA binding moieties [21]. Several activation domains have met this criterion which includes the 16-virion protein (VP16) of the herpes simplex virus type I, the VP64 domain is a tetramer of the minimal activator VP16, p65, the largest subunit of the NF-kappa B transcription factor, the Rta domain encoded by Kaposi's sarcoma-associated herpesvirus/human

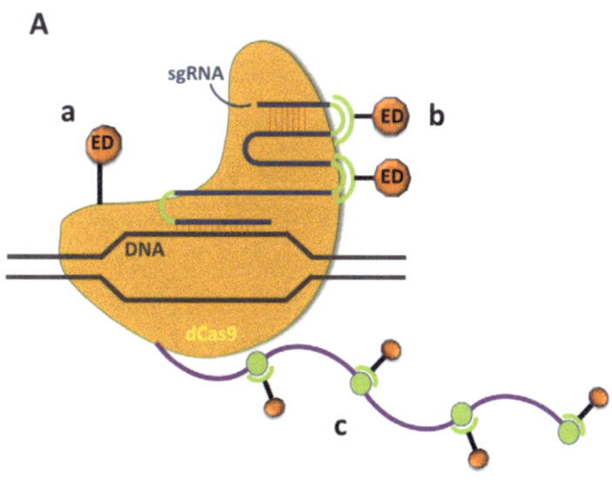


**Figure 1.**
*Strategies for gene activation and repression using the CRISPR Cas9 system. (A) Targeting specific DNA sequence for activation or repression through (a) direct binding of effector domain (ED) to dCas9; (b) effector domains bound to the sgRNA scaffold named srRNA 2.0; and (c) recruitment of the effector domain to the target site by the Sun tag system. (B) The concept of gene inactivation through CRISPR interference (CRISPRi) using KRAB bound to dCas9.*

herpesvirus 8, human heat shock factor 1 (HSF1) etc. [21]. The CRISPR activation cabinet (CRISPRa) consisting of dCas9 fused to the activation domain in various combinations has been used in the transcriptional activation of genes. In vitro studies showed that the dCas9-VP64 and dCas9-p65 fusion proteins have activated the target reporter gene in the HEK293 reporter cell line [22]. However, the VPR tripartite activation domain designed and characterized by George et al. reported a robust multi-locus activation, outpacing VP64 alone. Furthermore, for nuclease and activation pursuit using the dCas9-VPR system, maximum efficiency was achieved by changing the length of the gRNAs [23].

#### *2.1.1 Increment in the transactivation potential using dCas9*

Several interesting CRISPRa studies have described successful gene activation using dCas9-based tools and increased transcriptional activation with multiple targeting of sgRNA. However, Konermann et al. [24] have introduced the concept of increased activation using single sgRNA and modification of the gRNA scaffold and have named it sgRNA2.0. The main objective was to create a flexible sgRNA capable of grafting additional activation domains to the target site dCas9. Enhanced stimulation can be achieved with a single sgRNA through multiple activation domains. A similar concept of activation domain recruitment to the target site was developed by Tanenbaum et al. [25] and named the Sun tag system. This approach uses a recombinant peptide array (SunTag) that recruits multiple antibody fusion proteins (activator domains). This system achieved higher levels of transcription activation with recruitment of VP64 to dCas9 through Suntag to activate the CXCR4 gene in K562 cell lines using a single sgRNA. It achieved 25 times higher transcription levels than conventional dCas9-VP64-mediated activation [25]. In addition to the techniques manipulated by dCas9 to increase gene expression activation, light-controlled manipulation of epigenetic and endogenous genes was introduced [24, 26, 27]. The principle involved in this mode of gene manipulation is that two interaction partners dimerize upon blue-light activation, where one partner is fused to a DNA-binding protein and the other to the transcriptional activation domain. Recently, optogenetic systems have used dCas9 and photoinducible dimerization for precise modulation of gene expression of mammalian cells.

#### *2.1.2 dCas9-mediated transcriptional repression*

Qi et al. initially developed the concept of CRISPR interference (CRISPRi) [12], where dCas9 and sg RNAs target a specific transcriptional start site and interfere with the activity of DNA binding proteins such as Polymerase II. With this development of CRISPRi, modifications have been made in which a stronger repressor complex, such as Kruppel's associated Box (KRAB), was fused to dCas9, resulting in a stronger and more specific gene repressor than dCas9 alone [22]. KRAB has a highly conserved amino-terminal region of many Krüppel-class Cys2His2 zinc finger proteins with repressive function [28]. The KRAB domain functions with the KAP1 corepressor that recruits heterochromatin protein 1 (HP1), which recognizes and binds methylated H3K9, obtained by methyltransferase activity and recruits more HMTs, resulting in chromatin condensation of neighboring nucleosomes [29]. Several success stories reported the use of dCas9-KRAB fusion in gene repression, and this fusion system can successfully recruit chromatin-modifying complexes for amplified CRISPRi effects [21].

## 2.2 Epigenome editing through chromatin modulation using CRISPR Cas9

Genomic DNA is organized on a nucleosomal scale by wrapping around various histones that can be modified post-translationally using chemical moieties and/or by chemical modification, such as methylation at 5-carbon in cysteine residues (5-mC), which are collectively termed chromatin markers [2]. Gene editing through chromatin edits has become an important aspect in the epigenetic field in altering gene expression. dCas9 has been deployed to modulate various histone proteins within a specific DNA region that affects gene expression [2]. Although gene expression levels through chromatin modulation are modest compared to CRISPRi/a technologies, the highest levels of expression are achieved by H3K27ac deposition or DNA demethylation [30, 31].

### *2.2.1 dCas9 in methylation and demethylation of chromatin marks*

Since aberrant methylation has several pathological implications, especially in several cancers, there is an urgent need to manipulate these epigenetic characteristics. Although small molecules such as 5-azacytidine that target DNA methylation as epigenetic inhibitors are in clinical use, these inhibitors target the entire genome where normal methylation occurs [20]. This limitation has been overcome by targeting specific gene loci mediated by the dCas9 system, which has been used to deposit and remove methylation from target genome sites. The deposition of DNA methylation has been achieved by fusion of dCas9 with the catalytic domain of eukaryotic DNA methyl transferase (DNMT3A) [32, 33] or prokaryotic DNA methyltransferase (MQ3) [34]. In addition to targeted DNA methylation, the active removal of methylation marks from the endogenous DNA sequence is another mode of gene expression manipulation. Ten–eleven translocation (TET) proteins (TET1, TET2 and TET3) are involved in endogenous DNA demethylation that regulates cell type-specific gene expression. Therefore, in epigenetic DNA demethylation using the CRISPR system, guidable dCas9 fused to TET-1 has been used to achieve locus-specific DNA demethylation [31, 35–37].

### *2.2.2 dCas9 mediated modification of histone proteins*

In addition to the role of methylation and demethylation in gene regulation, epigenetic information is stored in histone proteins that wrap DNA around them to form chromatin fiber. These epigenetic features in a cell are constituted by post-translational modification of the histone tails that reveal key insights into regulatory activity. For example, mono and dimethylation at the four positions of lysine of Histone H3 (H3K4me1/2) and acetylation at the 27 positions of acetylation of Lysine (H3K27ac) are elements of active distal regulation. At the same time, trimethylation (H3K4me3) is a marker of active or poised bivalent promoters [11]. Since specific writers, correctors, and erasers regulate the modification of chromatin, researchers have used the dCas9 system to recruit histone modifiers for the methylation and demethylation of histone proteins [20]. Examples of histone-targeted epigenome editing include LSD1, a histone demethylase that removes the H3K4me2 mark, which was fused to dCas9, resulting in a substantial local decrease in the active enhancer markers H3K4me2 and H3K27ac and altered expression of target genes [38]. Unlike these findings, a fusion of histone acetyltransferase P300 with dCas9 resulted in a significant increase in local induction of H3K27ac expression in both the promoter and enhancer regions [39]. Histone modifications through the CRISPR system were used to suppress gene activity where local induction of H3K4me3, a marker of the active promoter for the re-expression of silenced targets, was used. Histone deacetylation by fusion of full-length histone deacetylases (HDAC) with dCas9 was shown to reduce H3k27ac and ultimately down-regulate target gene expression [40].

## 3. Gut microbiota and probiotics: the mysterious players in epigenetic changes

The gut microbiota/microbiome is the collective term used for the genes contributed collectively by viruses, bacteria, and fungi that are present in different regions of the human body and play a dynamic role in the health of the individual human [41]. Human gut co-habitants with various symbiotic microorganisms, including Bacteroides, Firmicutes, Actinobacteria, Proteobacteria, Fusobacteria, and Verrucomicrobia. The number and composition of these microbial species can vary according to age, diet, and lifestyle [42, 43]. The gut microbiota colonizes from the time of delivery and during breast feeding that gets continuously modulated during development due to dietary and environmental factors [44].

Food and Drug Administration (FDA) and World Health Organization (WHO) define probiotics as "live microorganisms that can provide health benefits to the host when administered in adequate amounts" [45]. Well-established probiotic bacteria include *Lactobacillus rhamnosus*, *Lactobacillus reuteri*, *Bifidobacteria*, *Lactobacillus casei*, *Lactobacillus acidophilus*, *Bacillus coagulans*, *Escherichia coli* strain Nissle 1917, and *Enterococcus faecium* SF68 [46]. In addition to protecting the intestinal tract, probiotics help improve various diseases, including cardiovascular disease, diabetes, and obesity [10, 47]. The well-known fact is that probiotics that restore intestinal microbiota balance support their efficiency in preventing the development of chronic immune-mediated diseases (**Figure 2**).

As there was substantial evidence that the intestinal microbiota and probiotics impact several metabolic functions within the human body, such as digestion, absorption of nutrition, regulation of hormonal secretion, modulation of inflammation and immunity processes, synthesis of vitamins and various metabolites, dysbiosis of this microbiota negatively influences the onset of diseases such as obesity, diabetes, inflammatory bowel disease, cancer, etc. [44, 48, 49]. In recent years, the commensal microbiota has also been appreciated to play a prominent role in host epigenetics by altering gene expression, affecting homeostasis and health status [44, 50, 51]. Although the mechanism through which the gut microbiota regulates host epigenetic changes remains a mystery, this has, however, been revealed through studies with advanced computational tools for metagenomics that have focused on understanding the crosstalk between the microbiota and epigenetics [52].

### 3.1 Gut microbiota and host epigenetic features

Host epigenetics can influence the shaping of the microbiota in the intestine. Studies have shown that host miRNAs could control microbial gene transcription, thus modeling the structure of the microbial community, vice versa was observed where the gut microbiota and its metabolites were found to regulate epigenetic changes and influence host metabolism [41]. Short-chain fatty acids (SCFAs) are the main metabolites in host epigenome modulation that are produced by intestinal

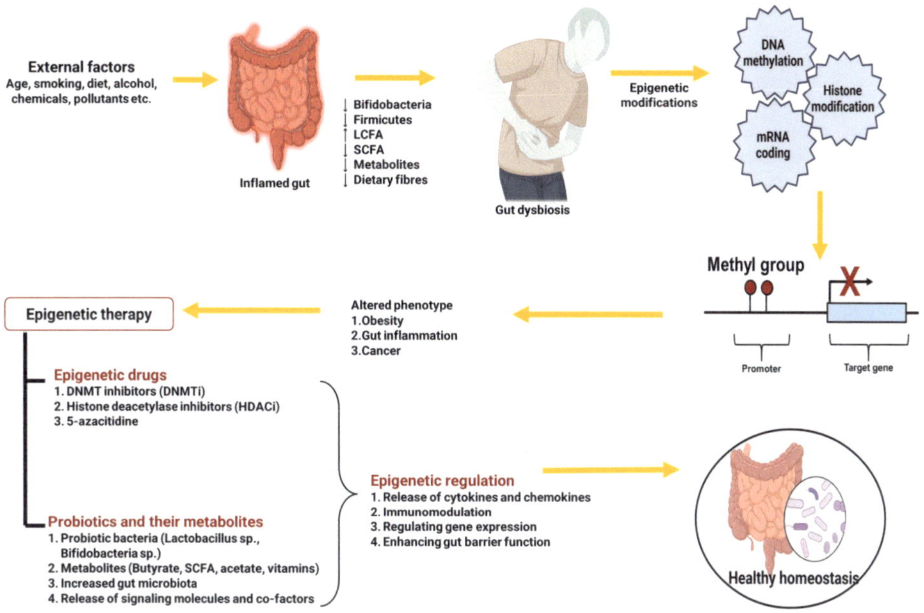


**Figure 2.**
*A schematic representation of the external factors responsible for inflamed gut and the altered phenotype due to epigenetic modifications. The figure also depicts epigenetic therapy and the restoration of a healthy gut by epigenetic drugs, probiotics, and their metabolites.*

microbes and are found to affect body health and disease status [53, 54]. The gut microbiota synthesizes many bioactive compounds that serve as epigenetic substrates, cofactors, or regulators of epigenetic enzyme activity [55, 56]. For example, folate and B vitamins (B2, B12) donate methyl groups for DNA or histone methylation. SCFAs inhibit the deacetylase activity of histone deacetylases (HDACs), leading to changes in chromatin [57, 58]. Studies in rodent models have shown that the gut microbiota was found to influence host epigenetic transcriptome markers. Alteration of intestinal microflora in normal, modified, and germ-free (GF) mice was found to affect the modification of N6-methyladenosine (m6A) in the cecum and the decrease in expression in the liver, which ultimately affected host metabolism, inflammation, and antibacterial process. *Akkermansia muciniphila* and *Lactobacillus plantarum* could influence the modification of m6A in mono-associated mice [59]. This microbiota was also found to modulate host miRNA expression patterns to maintain intestinal homeostasis [60]. The intestinal microflora negatively regulates miR-107 expression in dendritic cells and macrophages, thus affecting the activity of the MyD88 and NF-B pathways, which subsequently influenced immune homeostasis and the expression of the target gene IL-23p19, which are responsible for oral cancer [61]. Probiotic supplementation could prevent and treat colon cancer by regulating miRNAs [6]. Coculture of probiotic *Leuconostoc mesenteroides* with HT-29 demonstrated effective down-regulation of miRNA-21 and miRNA-200b, thus promoting colon cancer cell apoptosis [62].

### 3.2 Probiotic and host epigenetics

Probiotics reduce the risk of certain diseases primarily through clinically validated mechanisms involving epigenetic regulation. The intestinal flora in the large intestine

| Bacteria | Epigenetic Changes | Health effect | Reference |
|---|---|---|---|
| Lactobacilli | Down-regulated miRNAs (miR-200b, miR-215, and miR-192) | Maintain homeostasis and shape the host response to infection | [70] |
| *Leuconostoc mesenteroides* | Modulates miRNA-21, miRNA-200b | Promote the apoptosis of colon cancer cells | [62] |
| *Fusobacterium nucleatum* | Modulates the Toll-like receptor, miRNAs (miR-4802 and miR-18a) | Improve the response to chemotherapy, reduce cancer recurrence, | [71] |
| VSL#3 (VSL Pharmaceuticals, Ft Lauderdale, FL, USA): 3 × 1011 CFU/g of bifidobacteria (*B. longum*, *B. infantis*, and *B. breve*), lactobacilli (*L. acidophilus*, *L. casei*, *L. delbrueckii* subsp. *L. bulgaricus* and *L. plantarum*) and *Streptococcus salivarius* subsp. *thermophilus*. | Increases IL-10 production from Peyer patches and the spleen, along with increased expression of IL-10 in the pancreas. | Prevent the development of autoimmune diabetes in NOD mice and Induced immunomodulation by a reduction in insulitis severity | [72] |
| *Lactobacillus johnsonii* N6.2-Mediated | Increases the cytokines IL-6 and IL-23 within the mesenteric lymph nodes of BBDP fed LjN6.2. | Confirms the resistance of T1D | [73] |
| *Lactococcus lactis* | Increases the frequencies of local Tregs accumulated in the pancreatic islets | Restorative therapy for autoimmune diabetes | [74] |
| *Bifidobacterium* spp. | Reduces the expression of insulin receptor substrate 1, protein kinase B (Akt / PBB), IKKa and IkBa, protein-1 (MCP-1) and interleukin-6, (IL-6) | Promotes recovery of β-cells of pancreas cell-cells and increases insulin sensitivity in mice by enhancing the function of the insulin signaling pathway as a promising strategy for the treatment of diabetes. | [75] |
| *Lactobacillus reuteri* | inhibits osteoblast TNF-a signaling and osteoblast maturation markers. | Blocks the loss of bones | [76] |
| *Lactobacillus kefiranofaciens* M and *Lactobacillus kefiri* K | Stimulates GLP-1 secretion, inhibits pro-inflammatory and inflammatory cytokine production, increases IL-10 production, and modifies the intestinal microbiota cytokine TNFa and TH | Potentially inhibits type 1 diabetes progression in vitro and in vivo by improving GLP-1 | [77] |
| *Lactobacillus brevis* KLDS1.0727 and KLDS 1.0373 | GABA overexpression | Inhibits the development of T1D in diabetics mice model | [78] |
| *Lactobacillus johnsonii* N6.2 | decreases the K: T ratio significantly (CD45RO + CD183 + CD196–) and cytotoxic CD8+ T cells. | Reduced risk of T1D | [79] |

| Bacteria | Epigenetic Changes | Health effect | Reference |
|---|---|---|---|
| *Lactobacillus rhamnosus* GG and *Bifidobacterium lactis* Bb12 | Prevents islet cell destruction | Reduce the risk of autoimmunity | [80] |

**Table 1.**
*Probiotic bacteria with their epigenetic changes.*

undergoes fermentation to transform dietary fiber into SCFAs, with butyrate being one of the major SCFAs produced [41]. SCFAs such as acetate, propionate, and butyrate are found to modulate the host's immune system [63]. Gut bacteria and some probiotics produce enzymes such as methyltransferases, acetyltransferases, deacetylases, Bir A ligase, phosphotransferases, kinases, and synthetases. Furthermore, they produce S-adenosylmethionine (SAM), acetyl-CoA, NAD+, α-KG, and ATP, which are essential cofactors for many epigenetic processes that regulate DNA methylation, post-translational histone modifications, and nucleosome position [41, 44, 51]. Butyrate is an anti-inflammatory molecule produced by the gut microbiota and some probiotics can influence the host's epigenome by decreasing intestinal permeability, preventing autoimmunity for type 1 diabetes (T1D), and protecting against autoimmunity of islets [64, 65].

SCFAs are the main probiotic metabolites that have been indicated to have a protective function in the intestine and exert anti-inflammatory effects in several animal and human models. SCFAs exert health-promoting actions by lowering intestinal pH, acting as energy sources for colon-inhabiting microbes, stimulating colonic blood flow, helping in the contraction of smooth muscle cells, enhancing transepithelial chloride secretion, and aiding in the proliferation of colonic epithelial cells. These SCFAs are the class of epigenetic drugs with histone deacetylase inhibitory activity (HDACi) that plays a vital role as anticancer agents with strong antiproliferative effects on tumor cells [66, 67]. The administration of SCFAs, such as butyrate and acetate, has shown promising results in ameliorating inflammatory lesions in mouse models of allergic airway disease and colitis [68]. The probiotic bacterium *Propionibacterium freudenreichii* produced a high level of SCFA, acetate, and propionate [69]. Butyrate, acetate, polyphenols, and vitamins are the main metabolites of intestinal microbes that participate in epigenetic processes. **Table 1** lists some examples of probiotic bacteria that modulate the epigenetic makeup with their functional attributes. Studies have shown that butyrate can reduce inflammation (by increasing IL-10 expression) in inflammatory bowel disease and help protect against colitis and mortality [81]. Another LMW associated with epigenetic regulation is acetate, which is one of the immunomodulatory peptide drugs, "Glatiramer Acetate". This was reported to increase the expression of the FOXP3 transcription factor (Forkhead box P3), which increased Treg cell proliferation and differentiation. This reduced the rate of diabetes and insulitis in NOD mice [82]. These SCFAs inhibit HDAC activity, protect against neurodegenerative diseases, and promote immunoglobulin secretion and intestinal mucosal barrier function [83]. Vitamins such as B2, B12, and B6 are essential for an enzyme that systemizes SA, the primary methyl-donating substrate for DNMT and HHMT. Probiotic species such as *Lactobacillus* and *Bifidobacterium* can confer better protective effects by biotinylating proteins in the intestines. This process is important for DNA repair and chromatin structure [84].

## 4. Epigenetic therapies

Epigenetic therapies are a promising new area of treatment for cancer and other diseases. They involve manipulating gene expression using drugs, probiotic bacteria, enzymes, or other molecules that interact with DNA. The gut microbiota and probiotics can affect the host epigenome by activating epigenetically silenced genes in cancer cells. Butyrate, produced by gut microbiota can prevent colon cancer and other diseases by reducing pro-angiogenic factors such as EGF and HIF 1α [85]. Sodium butyrate is an HDAC inhibitor and can increase cell death in human medulloblastoma cells [86]. Below we have discussed the epigenetic treatment for major organs that are effected by cancer, including lung, liver, gastric colon, breast, etc.

### 4.1 Epigenetic therapy in gastric cancer (GC)

GC is one of the most fatal forms of cancer in the world. Its grim prognosis is due to the complexity of the disease, late diagnosis, and unsatisfactory treatments. In addition to genetic changes and external influences, studies have shown that modifications in epigenetic processes are instrumental in the emergence and development of gastric tumors, which has become a distinctive feature of GC [87]. In most cases, it is caused by *Helicobacter pylori* infection characterized by chronic gastritis and peptic ulcers. *H. pylori* harboring the pathogenicity island of the cytotoxins-associated gene (CagPAI) induces the dephosphorylation of histone H3S10, H3 threonine 3, and the deacetylation of H3K23 in gastric epithelial cells [88]. DNMT inhibitors (DNMTi) and histone deacetylase inhibitors (HDACi) have been shown to be the most effective epigenetic drugs for treating GC in animal models. Nucleoside analogues of DNMTi (such as 5-azacitidine and 5-aza-dC or decitabine (DAC)) and non-nucleoside analogues of DNMTi (such as hydralazine) are separated according to their ability to integrate into freshly produced DNA [89]. A total of 485,512 methylation locations (482,421 at CpG sites and 3091 at non-CpG sites) from 55 cancer-related genes showed that epigenetic aberrations could alter various cancer-related pathways [87]. With the addition of the epigenetic drug to neoadjuvant epirubicin-oxaliplatin-capecitabine, a response rate of 67% was observed in patients with well-differentiated gastro-esophageal tumors, including 25% with a complete response. Hypomethylation of tumor-associated loci was observed at all 5-azacitidine doses, and the degrees of hypomethylation were correlated with a therapeutic response [90]. Gut bacteria can help fight cancer by producing bioactive metabolites that inhibit HDAC and HAT. These metabolites like SCFAs, propionate and butyrate can alter the epigenetic profile of cancer cells, making them more susceptible to treatment [91]. However, changing epigenetic profiles may represent new treatment approaches for GC. Cancer heterogeneity can be overcome and cancer homeostasis can be reprogrammed so that it is more likely to react to cytotoxic drugs or immune checkpoint inhibitors if epigenetic processes are targeted.

### 4.2 Epigenetic therapy in colon cancer

Colon cancer is one of the most fatal forms of the disease worldwide. Its genesis is linked to the collection of genetic and epigenetic modifications in the epithelial cells of the colon that cause them to morph into adenocarcinomas. In the last 10 years, there has been enormous progress in understanding cancer epigenetics, notably aberrant DNA methylation [92]. Microbial exposure was perceived to induce DNA hypomethylation, which increases the expression and activity of the DNA demethylase enzymes Tet3 and Dnmt1 [50, 93]. DNA methylation in colon biopsies is correlated with microbial composition, inflammation status, and disease classification in ulcerative colitis and Crohn's disease patients [85]. Ulcerative colitis is associated with colorectal cancer, since *Fusobacterium* is associated with increased DNA methylation [94–96]. Yu et al. [71] demonstrated DNA methylation, particularly in the 3'CpG islands of glycosylation genes involved in cell maturation, which was substantially hampered in the absence of the intestinal microbiota. Changes in the fecal microbiota with decreased SCFA (acetate, propionate, and butyrate) have been attributed to the development and progression of colon cancer [97]. Studies have shown a low level of butyrate-producing bacteria in patients with colorectal cancer with a concurrent increase in mucin-degrading species such as *Akkermansia muciniphila* [98]. Probiotics can reduce the risk of colon cancer by improving intestinal microbiota balance. They can also reduce side effects of chemotherapy and radiation therapy by regulating neutrophil function and supporting anti-inflammatory activity by inducing TNF-α and miRNA-dependent expression of p21 gene [99, 100].

### 4.3 Epigenetic therapy in breast, ovarian, and endometrial cancer

Most ovarian cancers, 14–24%, are inherited diseases caused by gene mutations in BRCA1 and BRCA2 [101]. When BRCA1, BRCA2, BRIP1, RAD51C, RAD51D, and FANCM are altered, genomic instability results in ovarian cancer with an early somatic mutation in TP53 [102]. The epigenetic environment of these tumors can contribute to increased immune activity. Although durable and long-lasting responses have been shown in solid tumors such as melanoma, lung cancer, and renal cell carcinoma [103], checkpoint blockade therapies have not been successful in ovarian cancer. Six genes that help control the cell cycle—BRCA1, CDKN2A, RASSF1A, LOTI, DAPK, and ICAM-l—are suppressed when hypermethylated gene promoters are detected in cancer. Studies have shown that extensive cessation of CpG hypermethylation in ovarian cancer leads to slower growth of cancer cells [104]. The research found that ITF2357 and 5-azacytidine (AZA) inhibited the DNMT1 enzyme and stimulated T cell participation by upregulating ERV expression in a mouse model of ovarian tumor. This activated a type I interferon response by increasing the expression of endogenous retroviruses (ERV), pieces of ancient viral DNA that make up 8% of our genome [105]. Moufarrij et al. [106] demonstrated that DNMTi decitabine successfully treats ovarian cancer that does not respond to chemotherapy. Furthermore, Cicek et al. [107] determined that the epigenetic drugs can enhance expression of tumor antigen NY-ESO-1 by changing its methylation status. It is important to study the combined pharmacodynamics and multifactorial mechanisms of epigenetic drugs when combined with targeted or immune therapies.

The gut microbiome has been observed to modulate estrogen metabolism [108]. Shimizu et al. [109] showed a significant increase in the reproductive capacity of germ-free mice with the introduction of microbials. Bacterial introduction normalized the estrous cycle and increased the copulation and implantation rates. An aggregate of enteric bacterial genes in the human gut microbiome can influence the estrabolome. Bacterial β-glucuronidases and β-glucuronides enhance estrogen deconjugation and conjugation [110]. Deconjugation results in the re-absorption of free estrogens, leading to the development of estrogen-driven cancers such as breast, ovarian, and endometrial cancers. Daidzein, a class of hydroxy isoflavones, is converted to dihydrodaidzein, S-(−)equol, and O-desmethyl angiotensin by the concentration of equol in the urine of intestinal bacteria has been correlated with a reduced risk of breast cancer. Therefore, epigenetic modifications can affect cancer progression [111].

### 4.4 Epigenetic therapy in liver and lung cancer

Tumor forms of cancer are thought to result from oncogene activation and TSG silencing due to mutations in epigenetic regulatory pathways. These epigenetic alterations can also be related to resistance to chemotherapy [112]. Studies show that methylation of certain genes (like CDK2A, p16, CDH13, RASSF1A, and APC) correlates with the recurrence of stage I non-small cell lung cancer (NSCLC) after surgical resection and epigenetic changes that affect p16, and p16 expression are associated with reduced survival after early-stage NSCLC resection [113]. In a phase I/II trial, patients with advanced untreated NSCLC received an 8-hour continuous infusion of high-dose decitabine (200 to 660 mg/m2). Only one patient completed more than one cycle, affecting the efficacy of treatment [114]. Pharmacodynamic studies have found that one-third of patients with lung cancer have higher levels of p16, MAGE-3, and NY-ESO-1 [115]. Histone modifications, HAT, and HDAC inhibitors are currently used in lung cancer treatment, but are still a few years away from being used as lung cancer biomarkers and guide therapy [116].

Dietary regulation, balanced gut microflora, and epigenetic modification can be useful in preventing liver and lung cancer. Studies have shown that chromatin modifications and other epigenetics (especially miRNA) are critical to the development of chronic obstructive lung disease (COPD) and lung cancer [117, 118]. SCFAs produced by the gut microbiota bind to the G protein-coupled receptor 43 (GPR43), affecting inflammatory responses [30]. GPR43 is important in reducing inflammation in models of colitis, arthritis, and asthma [68].

## 5. Conclusions

As scientists continue to explore the potential applications of epigenetic reprogramming, the possibilities are endless. This research could transform the way a disease is treated and prevent it from occurring in the first place. There is much to learn about this process, but the medical and scientific communities are excited to explore the potential of these organisms. Epigenetic detection has been an important advance in cancer research. By shedding light on how cancer develops and progresses, new treatments have been developed on the basis of these findings. In spite of many challenges like aberrant promoter methylation, inactivated mutations, tumor microenvironment, and external factors, these developments could have a significant impact on the fight against cancer. Epigenetic therapies are a promising new field of medicine that has the potential to revolutionize healthcare. Epigenetic drugs and other therapies are being studied for their potential to treat a wide range of diseases, including cancer. The gut microbiome can be used to manage and treat cancer and gut-related diseases. The bioactive metabolites produced by beneficial probiotic microbes affect the epigenome and the differentiation and functioning of various immune cells, enterocytes, and pancreatic cells. Modifying the intestinal microbiome by repairing dysbiosis through diet intervention or fecal transplants would be a promising approach to treating metabolic syndromes. Additionally, several clinical trials of epigenetic therapy are currently underway and could offer insight into the potential of epigenetic therapies. A new generation of epigenetic therapies could revolutionize healthcare care and change the way we live in the near future.

## Acknowledgements

The authors are grateful to the editors for invitation of the book chapter. The authors SCR and SPR sincerely thank the founder of SVR BioScience Research Services, Dr. Sundru Manjulata Devi for giving an opportunity to contribute to the review.

## Conflict of interest

The authors declare no conflict of interest.

## Authorship criteria

Concept and design of the book chapter was conceived by SMD. The epigenetic changes and therapy was contributed by SMD. SCR contributed and formulated the for role of CRISR-Cas9 and SPR promoted the role of gut microbiota and probiotics in epigenetic changes. Editing, revision and final approval was made by SMD, SCR and SPR.

## Author details

Surya Chandra Rao Thumu[1], Shobha Rani Papanna[2] and Sundru Manjulata Devi[3]*

1 Centre for Neuroscience, Indian Institute of Science, Bangalore, Karnataka, India

2 Genespy Research Services Private Limited, Mysuru, Karnataka, India

3 SVR BioScience Research Services, Mysuru, Karnataka, India

*Address all correspondence to: svrbioscience@gmail.com; manjusundru@gmail.com

## References


[1] Morovic W, Budinoff CR. Epigenetics: A new frontier in probiotic research. Trends in Microbiology. 2021;**29**:117-126. DOI: 10.1016/j.tim.2020.04.008

[2] Nakamura M, Gao Y, Dominguez AA, Qi LS. CRISPR technologies for precise epigenome editing. Nature Cell Biology. 2021;**23**:11-22. DOI: 10.1038/s41556-020-00620-7

[3] Keung AJ, Joung JK, Khalil AS, Collins JJ. Chromatin regulation at the frontier of synthetic biology. Nature Reviews. Genetics. 2015;**16**:159-171. DOI: 10.1038/nrg3900

[4] Holtzman L, Gersbach CA. Editing the epigenome: Reshaping the genomic landscape. Annual Review of Genomics and Human Genetics. 2018;**19**:43-71. DOI: 10.1146/annurev-genom-083117-021632

[5] Gasperini M, Tome JM, Shendure J. Towards a comprehensive catalogue of validated and target-linked human enhancers. Nature Reviews. Genetics. 2020;**21**:292-310. DOI: 10.1038/s41576-019-0209-0

[6] Kumar M, Hemalatha R, Kumar R, Nagpal R, Devraj JP, Verma V, et al. Probiotics, prebiotics and synbiotics—A review. Cell. 2021;**8**:93-95. DOI: 10.4049/jimmunol.1001864

[7] Jones PA, Issa J-PJ, Baylin S. Targeting the cancer epigenome for therapy. Nature Reviews. Genetics. 2016;**17**:630-641. DOI: 10.1038/nrg.2016.93

[8] Wang Y, Yan L, Zhang Z, Prado E, Fu L, Xu X, et al. Epigenetic regulation and its therapeutic potential in pulmonary hypertension. Frontiers in Pharmacology. 2018;**9**:241. DOI: 10.3389/fphar.2018.00241

[9] Roberti A, Valdes AF, Torrecillas R, Fraga MF, Fernandez AF. Epigenetics in cancer therapy and nanomedicine. Clinical Epigenetics. 2019;**11**:81. DOI: 10.1186/s13148-019-0675-4

[10] Kesika P, Sivamaruthi BS, Chaiyasut C. Do probiotics improve the health status of individuals with diabetes mellitus? A review on outcomes of clinical trials. BioMed Research International. 2019;**2019**:1531567. DOI: 10.1155/2019/1531567

[11] Xiong X, Chen M, Lim WA, Zhao D, Qi LS. CRISPR/Cas9 for human genome engineering and disease research. Annual Review of Genomics and Human Genetics. 2016;**17**:131-154. DOI: 10.1146/annurev-genom-083115-022258

[12] Qi LS, Larson MH, Gilbert LA, Doudna JA, Weissman JS, Arkin AP, et al. Repurposing CRISPR as an RNA-guided platform for sequence-specific control of gene expression. Cell. 2013;**152**:1173-1183. DOI: 10.1016/j.cell.2013.02.022

[13] Morton AR, Dogan-Artun N, Faber ZJ, MacLeod G, Bartels CF, Piazza MS, et al. Functional enhancers shape extrachromosomal oncogene amplifications. Cell. 2019;**179**:1330-1341.e13. DOI: 10.1016/j.cell.2019.10.039

[14] Mirabella AC, Foster BM, Bartke T. Chromatin deregulation in disease. Chromosoma. 2016;**125**:75-93. DOI: 10.1007/s00412-015-0530-0

[15] Franzago M, Fraticelli F, Nicolucci A, Celentano C, Liberati M, Stuppia L, et al. Molecular analysis of a genetic variants panel related to

nutrients and metabolism: Association with susceptibility to gestational diabetes and Cardiometabolic risk in affected women. Journal Diabetes Research. 2017;**2017**:4612623. DOI: 10.1155/2017/4612623

[16] Hidalgo-Cantabrana C, O'Flaherty S, Barrangou R. CRISPR-based engineering of next-generation lactic acid bacteria. Current Opinion in Microbiology. 2017;**37**:79-87. DOI: 10.1016/j.mib.2017.05.015

[17] Pan M, Barrangou R. Combining omics technologies with CRISPR-based genome editing to study food microbes. Current Opinion in Biotechnology. 2020;**61**:198-208. DOI: 10.1016/j.copbio.2019.12.027

[18] Canani RB, di Costanzo M, Leone L, Bedogni G, Brambilla P, Cianfarani S, et al. Epigenetic mechanisms elicited by nutrition in early life. Nutrition Research Reviews. 2011;**24**:198-205. DOI: 10.1017/s0954422411000102

[19] Delgado S, Sánchez B, Margolles A, Ruas-Madiedo P, Ruiz L. Molecules produced by probiotics and intestinal microorganisms with immunomodulatory activity. Nutrients. 2020;**12**:391. DOI: 10.3390/nu12020391

[20] Adli M. The CRISPR tool kit for genome editing and beyond. Nature Communications. 2018;**9**:1911. DOI: 10.1038/s41467-018-04252-2

[21] Tadić V, Josipović G, Zoldoš V, Vojta A. CRISPR/Cas9-based epigenome editing: An overview of dCas9-based tools with special emphasis on off-target activity. Methods. 2019;**164-165**:109-119. DOI: 10.1016/j.ymeth.2019.05.003

[22] Gilbert LA, Larson MH, Morsut L, Liu Z, Brar GA, Torres SE, et al. CRISPR-mediated modular RNA-guided regulation of transcription in eukaryotes. Cell. 2013;**154**:442-451. DOI: 10.1016/j.cell.2013.06.044

[23] Kiani S, Chavez A, Tuttle M, Hall RN, Chari R, Ter-Ovanesyan D, et al. Cas9 gRNA engineering for genome editing, activation and repression. Nature Methods. 2015;**12**:1051-1054. DOI: 10.1038/nmeth.3580

[24] Konermann S, Brigham MD, Trevino A, Hsu PD, Heidenreich M, Cong L, et al. Optical control of mammalian endogenous transcription and epigenetic states. Nature. 2013;**500**:472-476. DOI: 10.1038/nature12466

[25] Tanenbaum ME, Gilbert LA, Qi LS, Weissman JS, Vale RD. A protein-tagging system for signal amplification in gene expression and fluorescence imaging. Cell. 2014;**159**:635-646. DOI: 10.1016/j.cell.2014.09.039

[26] Konermann S, Brigham MD, Trevino AE, Joung J, Abudayyeh OO, Barcena C, et al. Genome-scale transcriptional activation by an engineered CRISPR-Cas9 complex. Nature. 2015;**517**:583-588. DOI: 10.1038/nature14136

[27] Polstein LR, Gersbach CA. Light-inducible spatiotemporal control of gene activation by customizable zinc finger transcription factors. Journal of the American Chemical Society. 2012;**134**:16480-16483. DOI: 10.1021/ja3065667

[28] Margolin JF, Friedman JR, Meyer WK, Vissing H, Thiesen HJ, Rauscher FJ 3rd. Krüppel-associated boxes are potent transcriptional repression domains. Proceedings of the National Academy of Sciences of the United States of America. 1994;**91**:4509-4513. DOI: 10.1073/pnas.91.10.4509

[29] Fritsch L, Robin P, Mathieu JRR, Souidi M, Hinaux H, Rougeulle C, et al. A subset of the histone H3 lysine 9 methyltransferases Suv39h1, G9a, GLP, and SETDB1 participate in a multimeric complex. Molecular Cell. 2010;**37**:46-56. DOI: 10.1016/j.molcel.2009.12.017

[30] Lu Y, Chan Y-T, Tan H-Y, Li S, Wang N, Feng Y. Epigenetic regulation in human cancer: The potential role of epi-drug in cancer therapy. Molecular Cancer. 2020;**19**:79. DOI: 10.1186/s12943-020-01197-3

[31] Morita S, Noguchi H, Horii T, Nakabayashi K, Kimura M, Okamura K, et al. Targeted DNA demethylation in vivo using dCas9–peptide repeat and scFv–TET1 catalytic domain fusions. Nature Biotechnology. 2016;**34**:1060-1065. DOI: 10.1038/nbt.3658

[32] Amabile A, Migliara A, Capasso P, Biffi M, Cittaro D, Naldini L, et al. Inheritable silencing of endogenous genes by hit-and-run targeted epigenetic editing. Cell. 2016;**167**:219-232.e14. DOI: 10.1016/j.cell.2016.09.006

[33] Vojta A, Dobrinić P, Tadić V, Bočkor L, Korać P, Julg B, et al. Repurposing the CRISPR-Cas9 system for targeted DNA methylation. Nucleic Acids Research. 2016;**44**:5615-5628. DOI: 10.1093/nar/gkw159

[34] Lei Y, Zhang X, Su J, Jeong M, Gundry MC, Huang Y-H, et al. Targeted DNA methylation in vivo using an engineered dCas9-MQ1 fusion protein. Nature Communications. 2017;**8**:16026. DOI: 10.1038/ncomms16026

[35] Liu J, Yue Y, Han D, Wang X, Fu Y, Zhang L, et al. A METTL3-METTL14 complex mediates mammalian nuclear RNA N6-adenosine methylation. Nature Chemical Biology. 2014;**10**:93-95. DOI: 10.1038/nchembio.1432

[36] Xu X, Tao Y, Gao X, Zhang L, Li X, Zou W, et al. A CRISPR-based approach for targeted DNA demethylation. Cell Discovery. 2016;**2**:16009. DOI: 10.1038/celldisc.2016.9

[37] Choudhury SR, Cui Y, Lubecka K, Stefanska B, Irudayaraj J. CRISPR-dCas9 mediated TET1 targeting for selective DNA demethylation at BRCA1 promoter. Oncotarget. 2016;**7**:46545-46556. DOI: 10.18632/oncotarget.10234

[38] Kearns NA, Pham H, Tabak B, Genga RM, Silverstein NJ, Garber M, et al. Functional annotation of native enhancers with a Cas9-histone demethylase fusion. Nature Methods. 2015;**12**:401-403. DOI: 10.1038/nmeth.3325

[39] Hilton IB, D'Ippolito AM, Vockley CM, Thakore PI, Crawford GE, Reddy TE, et al. Epigenome editing by a CRISPR-Cas9-based acetyltransferase activates genes from promoters and enhancers. Nature Biotechnology. 2015;**33**:510-517. DOI: 10.1038/nbt.3199

[40] Kwon DY, Zhao Y-T, Lamonica JM, Zhou Z. Locus-specific histone deacetylation using a synthetic CRISPR-Cas9-based HDAC. Nature Communications. 2017;**8**:15315. DOI: 10.1038/ncomms15315

[41] Paul B, Barnes S, Demark-Wahnefried W, Morrow C, Salvador C, Skibola C, et al. Influences of diet and the gut microbiome on epigenetic modulation in cancer and other diseases. Clinical Epigenetics. 2015;**7**:112. DOI: 10.1186/s13148-015-0144-7

[42] Hollister EB, Gao C, Versalovic J. Compositional and functional features of the gastrointestinal microbiome and their effects on human health. Gastroenterology. 2014;**146**:1449-1458. DOI: 10.1053/j.gastro.2014.01.052

[43] Rinninella E, Raoul P, Cintoni M, Franceschi F, Miggiano GAD, Gasbarrini A, et al. What is the healthy gut microbiota composition? A changing ecosystem across age, environment, diet, and diseases. Microorganisms. 2019;**7**:14. DOI: 10.3390/microorganisms7010014

[44] Sharma M, Li Y, Stoll ML, Tollefsbol TO. The epigenetic connection between the gut microbiome in obesity and diabetes. Frontiers in Genetics. 2020;**10**:1329. DOI: 10.3389/fgene.2019.01329

[45] Hill C, Guarner F, Reid G, Gibson GR, Merenstein DJ, Pot B, et al. The international scientific Association for Probiotics and Prebiotics consensus statement on the scope and appropriate use of the term probiotic. Nature Reviews Gastroenterology & Hepatology. 2014;**11**:506-514. DOI: 10.1038/nrgastro.2014.66

[46] Pandey KR, Naik SR, Vakil B. Probiotics, prebiotics and synbiotics—A review. Journal of Food Science and Technology. 2015;**52**:7577-7587. DOI: 10.1007/s13197-015-1921-1

[47] Moludi J, Maleki V, Jafari-Vayghyan H, Vaghef-Mehrabany E, Alizadeh M. Metabolic endotoxemia and cardiovascular disease: A systematic review about potential roles of prebiotics and probiotics. Clinical and Experimental Pharmacology & Physiology. 2020;**47**:927-939. DOI: 10.1111/1440-1681.13250

[48] Cani PD, Everard A, Duparc T. Gut microbiota, enteroendocrine functions and metabolism. Current Opinion in Pharmacology. 2013;**13**:935-940. DOI: 10.1016/j.coph.2013.09.008

[49] Louis P, Hold GL, Flint HJ. The gut microbiota, bacterial metabolites and colorectal cancer. Nature Reviews. Microbiology. 2014;**12**:661-672. DOI: 10.1038/nrmicro3344

[50] Ansari I, Raddatz G, Gutekunst J, Ridnik M, Cohen D, Abu-Remaileh M, et al. The microbiota programs DNA methylation to control intestinal homeostasis and inflammation. Nature Microbiology. 2020;**5**:610-619. DOI: 10.1038/s41564-019-0659-3

[51] Sharma S, Kelly TK, Jones PA. Epigenetics in cancer. Carcinogenesis. 2010;**31**:27-36. DOI: 10.1093/carcin/bgp220

[52] Woo V, Alenghat T. Epigenetic regulation by gut microbiota. Gut Microbes. 2022;**14**:2022407. DOI: 10.1080/19490976.2021.2022407

[53] Wang Y, Kuang Z, Yu X, Ruhn KA, Kubo M, Hooper L, et al. The intestinal microbiota regulates body composition through NFIL3 and the circadian clock. Science. 2017;**357**:912-916. DOI: 10.1126/science.aan0677

[54] Cervantes-Barragan L, Chai JN, MaD T, di Luccia B, Ahern PP, Merriman J, et al. *Lactobacillus reuteri* induces gut intraepithelial CD4 $^+$ CD8αα $^+$ T cells. Science. 1979;**2017**(357):806-810. DOI: 10.1126/science.aah5825

[55] Miro-Blanch J, Yanes O. Epigenetic regulation at the interplay between gut microbiota and host metabolism. Frontiers in Genetics. 2019;**10**:638. DOI: 10.3389/fgene.2019.00638

[56] Shock T, Badang L, Ferguson B, Martinez-Guryn K. The interplay between diet, gut microbes, and host epigenetics in health and disease. The Journal of Nutritional Biochemistry. 2021;**95**:108631. DOI: 10.1016/j.jnutbio.2021.108631

[57] Arpaia N, Campbell C, Fan X, Dikiy S, van der Veeken J, deRoos P, et al.

Metabolites produced by commensal bacteria promote peripheral regulatory T-cell generation. Nature. 2013;**504**:451-455. DOI: 10.1038/nature12726

[58] Chang P, v, Hao L, Offermanns S, Medzhitov R. The microbial metabolite butyrate regulates intestinal macrophage function via histone deacetylase inhibition. Proceedings of the National Academy of Sciences of the United States of America. 2014;**111**:2247-2252. DOI: 10.1073/pnas.1322269111

[59] Jabs S, Biton A, Bécavin C, Nahori M-A, Ghozlane A, Pagliuso A, et al. Impact of the gut microbiota on the m(6) a epitranscriptome of mouse cecum and liver. Nature Communications. 2020;**11**:1344. DOI: 10.1038/s41467-020-15126-x

[60] Dalmasso G, Nguyen HTT, Yan Y, Laroui H, Charania MA, Ayyadurai S, et al. Microbiota modulate host gene expression via microRNAs. PLoS One. 2011;**6**:e19293-e19293. DOI: 10.1371/journal.pone.0019293

[61] Xue X, Cao AT, Cao X, Yao S, Carlsen ED, Soong L, et al. Downregulation of microRNA-107 in intestinal CD11c(+) myeloid cells in response to microbiota and proinflammatory cytokines increases IL-23p19 expression. European Journal of Immunology. 2014;**44**:673-682. DOI: 10.1002/eji.201343717

[62] Zununi Vahed S, Barzegari A, Rahbar Saadat Y, Goreyshi A, Omidi Y. Leuconostoc mesenteroides-derived anticancer pharmaceuticals hinder inflammation and cell survival in colon cancer cells by modulating NF-κB/AKT/PTEN/MAPK pathways. Biomedicine & Pharmacotherapy. 2017;**94**:1094-1100. DOI: 10.1016/j.biopha.2017.08.033

[63] Hullar MAJ, Fu BC. Diet, the gut microbiome, and epigenetics. Cancer Journal. 2014;**20**:170-175. DOI: 10.1097/PPO.0000000000000053

[64] Endesfelder D, Zu Castell W, Ardissone A, Davis-Richardson AG, Achenbach P, Hagen M, et al. Compromised gut microbiota networks in children with anti-islet cell autoimmunity. Diabetes. 2014;**63**:2006-2014. DOI: 10.2337/db13-1676

[65] Waldecker M, Kautenburger T, Daumann H, Busch C, Schrenk D. Inhibition of histone-deacetylase activity by short-chain fatty acids and some polyphenol metabolites formed in the colon. The Journal of Nutritional Biochemistry. 2008;**19**:587-593. DOI: 10.1016/j.jnutbio.2007.08.002

[66] Marks PA, Richon VM, Miller T, Kelly WK. Histone deacetylase inhibitors. Adv Cancer Res. 2004;**91**:137-168. DOI: 10.1016/S0065-230X(04)91004-4

[67] Acharya MR, Sparreboom A, Venitz J, Figg WD. Rational development of histone deacetylase inhibitors as anticancer agents: A review. Molecular Pharmacology. 2005;**68**:917-932. DOI: 10.1124/mol.105.014167

[68] Maslowski KM, Vieira AT, Ng A, Kranich J, Sierro F, Yu D, et al. Regulation of inflammatory responses by gut microbiota and chemoattractant receptor GPR43. Nature. 2009;**461**:1282-1286. DOI: 10.1038/nature08530

[69] Lan A, Lagadic-Gossmann D, Lemaire C, Brenner C, Jan G. Acidic extracellular pH shifts colorectal cancer cell death from apoptosis to necrosis upon exposure to propionate and acetate, major end-products of the human probiotic propionibacteria. Apoptosis. 2006;**12**:573-591. DOI: 10.1007/s10495-006-0010-3

[70] Wieërs G, Belkhir L, Enaud R, Leclercq S, Philippart de Foy J-M, Dequenne I, et al. How probiotics affect the microbiota. Frontiers in Cellular and Infection Microbiology. 2020;**9**:454. DOI: 10.3389/fcimb.2019.00454

[71] Yu T, Guo F, Yu Y, Sun T, Ma D, Han J, et al. Fusobacterium nucleatum promotes Chemoresistance to colorectal cancer by modulating autophagy. Cell. 2017;**170**:548-563.e16. DOI: 10.1016/j.cell.2017.07.008

[72] Calcinaro F, Dionisi S, Marinaro M, Candeloro P, Bonato V, Marzotti S, et al. Oral probiotic administration induces interleukin-10 production and prevents spontaneous autoimmune diabetes in the non-obese diabetic mouse. Diabetologia. 2005;**48**:1565-1575. DOI: 10.1007/s00125-005-1831-2

[73] Lau K, Benitez P, Ardissone A, Wilson TD, Collins EL, Lorca G, et al. Inhibition of type 1 diabetes correlated to a *lactobacillus johnsonii* N6.2-mediated Th17 bias. The Journal of Immunology. 2011;**186**:3538-3546. DOI: 10.4049/jimmunol.1001864

[74] Takiishi T, Korf H, van Belle TL, Robert S, Grieco FA, Caluwaerts S, et al. Reversal of autoimmune diabetes by restoration of antigen-specific tolerance using genetically modified Lactococcus lactis in mice. The Journal of Clinical Investigation. 2012;**122**:1717-1725. DOI: 10.1172/JCI60530

[75] Le TKC, Hosaka T, Nguyen TT, Kassu A, Dang TO, Tran HB, et al. Bifidobacterium species lower serum glucose, increase expressions of insulin signaling proteins, and improve adipokine profile in diabetic mice. Biomedical Research. 2015;**36**:63-70. DOI: 10.2220/biomedres.36.63

[76] Zhang J, Motyl KJ, Irwin R, MacDougald OA, Britton RA, McCabe LR. Loss of bone and Wnt10b expression in male type 1 diabetic mice is blocked by the probiotic *lactobacillus reuteri*. Endocrinology. 2015;**156**:3169-3182. DOI: 10.1210/EN.2015-1308

[77] Wei S-H, Chen Y-P, Chen M-J. Selecting probiotics with the abilities of enhancing GLP-1 to mitigate the progression of type 1 diabetes in vitro and in vivo. Journal of Functional Foods. 2015;**18**:473-486. DOI: 10.1016/j.jff.2015.08.016

[78] Abdelazez A, Abdelmotaal H, Evivie SE, Melak S, Jia F-F, Khoso MH, et al. Screening potential probiotic characteristics of lactobacillus brevis strains in vitro and intervention effect on type I diabetes in vivo. BioMed Research International. 2018;**2018**:7356173. DOI: 10.1155/2018/7356173

[79] Marcial GE, Ford AL, Haller MJ, Gezan SA, Harrison NA, Cai D, et al. Lactobacillus johnsonii N6.2 modulates the host immune responses: A double-blind, randomized trial in healthy adults. Frontiers in Immunology. 2017;**8**:655. DOI: 10.3389/fimmu.2017.00655

[80] Groele L, Szajewska H, Szypowska A. Effects of lactobacillus rhamnosus GG and Bifidobacterium lactis Bb12 on beta-cell function in children with newly diagnosed type 1 diabetes: Protocol of a randomised controlled trial. BMJ Open. 2017;**7**:e017178-e017178. DOI: 10.1136/bmjopen-2017-017178

[81] Sokol H, Seksik P, Furet JP, Firmesse O, Nion-Larmurier I, Beaugerie L, et al. Low counts of Faecalibacterium prausnitzii in colitis microbiota. Inflammatory Bowel Diseases. 2009;**15**:1183-1189. DOI: 10.1002/ibd.20903

[82] Cui J, Li F, Shi Z-L. Origin and evolution of pathogenic coronaviruses.

Nature Reviews. Microbiology. 2019;**17**:181-192. DOI: 10.1038/s41579-018-0118-9

[83] Kibe R, Kurihara S, Sakai Y, Suzuki H, Ooga T, Sawaki E, et al. Upregulation of colonic luminal polyamines produced by intestinal microbiota delays senescence in mice. Scientific Reports. 2014;**4**:4548. DOI: 10.1038/srep04548

[84] Dai Z, Zhang J, Wu Q, Chen J, Liu J, Wang L, et al. The role of microbiota in the development of colorectal cancer. International Journal of Cancer. 2019;**145**:2032-2041. DOI: 10.1002/ijc.32017

[85] Ryan FJ, Ahern AM, Fitzgerald RS, Laserna-Mendieta EJ, Power EM, Clooney AG, et al. Colonic microbiota is associated with inflammation and host epigenomic alterations in inflammatory bowel disease. Nature Communications. 2020;**11**:1512. DOI: 10.1038/s41467-020-15342-5

[86] Nör C, Sassi FA, de Farias CB, Schwartsmann G, Abujamra AL, Lenz G, et al. The histone deacetylase inhibitor sodium butyrate promotes cell death and differentiation and reduces Neurosphere formation in human Medulloblastoma cells. Molecular Neurobiology. 2013;**48**:533-543. DOI: 10.1007/s12035-013-8441-7

[87] Canale M, Casadei-Gardini A, Ulivi P, Arechederra M, Berasain C, Lollini P-L, et al. Epigenetic mechanisms in gastric cancer: Potential new therapeutic opportunities. International Journal of Molecular Sciences. 2020;**21**:5500. DOI: 10.3390/ijms21155500

[88] Ding S-Z, Goldberg JB, Hatakeyama M. Helicobacter pylori infection, oncogenic pathways and epigenetic mechanisms in gastric carcinogenesis. Future Oncology. 2010;**6**:851-862. DOI: 10.2217/fon.10.37

[89] Zhou Z, Li H-Q, Liu F. DNA methyltransferase inhibitors and their therapeutic potential. Current Topics in Medicinal Chemistry. 2019;**18**:2448-2457. DOI: 10.2174/1568026619666181120150122

[90] Schneider BJ, Shah MA, Klute K, Ocean A, Popa E, Altorki N, et al. Phase I study of epigenetic priming with Azacitidine prior to standard neoadjuvant chemotherapy for patients with Resectable gastric and esophageal adenocarcinoma: Evidence of tumor Hypomethylation as an indicator of major histopathologic response. Clinical Cancer Research. 2017;**23**:2673-2680. DOI: 10.1158/1078-0432.CCR-16-1896

[91] Sun D, Chen Y, Fang J-Y. Influence of the microbiota on epigenetics in colorectal cancer. National Science Review. 2019;**6**:1138-1148. DOI: 10.1093/nsr/nwy160

[92] Lao VV, Grady WM. Epigenetics and colorectal cancer. Nature Reviews. Gastroenterology & Hepatology. 2011;**8**:686-700. DOI: 10.1038/nrgastro.2011.173

[93] Poupeau A, Garde C, Sulek K, Citirikkaya K, Treebak JT, Arumugam M, et al. Genes controlling the activation of natural killer lymphocytes are epigenetically remodeled in intestinal cells from germ-free mice. The FASEB Journal. 2019;**33**:2719-2731. DOI: 10.1096/fj.201800787R

[94] Eaden JA, Abrams KR, Mayberry JF. The risk of colorectal cancer in ulcerative colitis: A meta-analysis. Gut. 2001;**48**:526-535. DOI: 10.1136/gut.48.4.526

[95] Tahara T, Hirata I, Nakano N, Tahara S, Horiguchi N, Kawamura T, et al.

Potential link between fusobacterium enrichment and DNA methylation accumulation in the inflammatory colonic mucosa in ulcerative colitis. Oncotarget. 2017;**8**:61917-61926. DOI: 10.18632/oncotarget.18716

[96] Linson EA, Hanauer SB. Epidemiology of colorectal cancer in inflammatory bowel disease—The evolving landscape. Current Gastroenterology Reports. 2021;**23**(9):1-16. DOI: 10.1007/s11894-021-00816-3

[97] Ohigashi S, Sudo K, Kobayashi D, Takahashi O, Takahashi T, Asahara T, et al. Changes of the intestinal microbiota, short chain fatty acids, and fecal pH in patients with colorectal cancer. Digestive Diseases and Sciences. 2013;**58**:1717-1726. DOI: 10.1007/s10620-012-2526-4

[98] Weir TL, Manter DK, Sheflin AM, Barnett BA, Heuberger AL, Ryan EP. Stool microbiome and metabolome differences between colorectal cancer patients and healthy adults. PLoS One. 2013;**8**:e70803-e70803. DOI: 10.1371/journal.pone.0070803

[99] Hýžďalová M, Hofmanová J, Pacherník J, Vaculová A, Kozubík A. The interaction of butyrate with TNF-α during differentiation and apoptosis of colon epithelial cells: Role of NF-κB activation. Cytokine. 2008;**44**:33-43. DOI: 10.1016/j.cyto.2008.06.003

[100] Hu S, Dong TS, Dalal SR, Wu F, Bissonnette M, Kwon JH, et al. The microbe-derived short chain fatty acid butyrate targets miRNA-dependent p21 gene expression in human colon cancer. PLoS One. 2011;**6**:e16221-e16221. DOI: 10.1371/journal.pone.0016221

[101] Walsh T, Casadei S, Lee MK, Pennil CC, Nord AS, Thornton AM, et al. Mutations in 12 genes for inherited ovarian, fallopian tube, and peritoneal carcinoma identified by massively parallel sequencing. Proceedings of the National Academy of Sciences of the United States of America. 2011;**108**:18032-18037. DOI: 10.1073/pnas.1115052108

[102] Kroeger PT Jr, Drapkin R. Pathogenesis and heterogeneity of ovarian cancer. Current Opinion in Obstetrics & Gynecology. 2017;**29**:26-34. DOI: 10.1097/GCO.0000000000000340

[103] Jelinic P, Ricca J, van Oudenhove E, Olvera N, Merghoub T, Levine DA, et al. Immune-active microenvironment in small cell carcinoma of the ovary, hypercalcemic type: Rationale for immune checkpoint blockade. Journal of the National Cancer Institute. 2018;**110**:787-790. DOI: 10.1093/jnci/djx277

[104] Ahluwalia A, Yan P, Hurteau JA, Bigsby RM, Jung SH, Huang TH-M, et al. DNA methylation and ovarian cancer. Gynecologic Oncology. 2001;**82**:261-268. DOI: 10.1006/gyno.2001.6291

[105] Stone ML, Chiappinelli KB, Li H, Murphy LM, Travers ME, Topper MJ, et al. Epigenetic therapy activates type I interferon signaling in murine ovarian cancer to reduce immunosuppression and tumor burden. Proceedings of the National Academy of Sciences of the United States of America. 2017;**114**:E10981-E10990. DOI: 10.1073/pnas.1712514114

[106] Moufarrij S, Dandapani M, Arthofer E, Gomez S, Srivastava A, Lopez-Acevedo M, et al. Epigenetic therapy for ovarian cancer: Promise and progress. Clinical Epigenetics. 2019;**11**:7. DOI: 10.1186/s13148-018-0602-0

[107] Cicek MS, Koestler DC, Fridley BL, Kalli KR, Armasu SM, Larson MC,

et al. Epigenome-wide ovarian cancer analysis identifies a methylation profile differentiating clear-cell histology with epigenetic silencing of the HERG K+ channel. Human Molecular Genetics. 2013;**22**:3038-3047. DOI: 10.1093/hmg/ddt160

[108] Fuhrman BJ, Feigelson HS, Flores R, Gail MH, Xu X, Ravel J, et al. Associations of the fecal microbiome with urinary estrogens and estrogen metabolites in postmenopausal women. The Journal of Clinical Endocrinology and Metabolism. 2014;**99**:4632-4640. DOI: 10.1210/jc.2014-2222

[109] Shimizu K, Muranaka Y, Fujimura R, Ishida H, Tazume S, Shimamura T. Normalization of reproductive function in germfree mice following bacterial contamination. Experimental Animals. 1998;**47**:151-158. DOI: 10.1538/expanim.47.151

[110] Plottel CS, Blaser MJ. Microbiome and malignancy. Cell Host & Microbe. 2011;**10**:324-335. DOI: 10.1016/j.chom.2011.10.003

[111] Rowland IR, Wiseman H, Sanders TAB, Adlercreutz H, Bowey EA. Interindividual variation in metabolism of soy Isoflavones and Lignans: Influence of habitual diet on Equol production by the gut microflora. Nutrition and Cancer. 2000;**36**:27-32. DOI: 10.1207/s15327914nc3601_5

[112] Jones PA, Baylin SB. The fundamental role of epigenetic events in cancer. Nature Reviews. Genetics. 2002;**3**:415-428. DOI: 10.1038/nrg816

[113] Brock M, Hooker CM, Ota-Machida E, Han Y, Guo M, Ames S, et al. DNA methylation markers and early recurrence in stage I lung cancer. New England Journal of Medicine. 2008;**358**:1118-1128. DOI: 10.1056/nejmoa0706550

[114] Momparler RL, Ayoub J. Potential of 5-aza-2′-deoxycytidine (Decitabine) a potent inhibitor of DNA methylation for therapy of advanced non-small cell lung cancer. Lung Cancer. 2001;**34**:111-115. DOI: 10.1016/s0169-5002(01)00397-x

[115] Schrump DS, Fischette MR, Nguyen DM, Zhao M, Li X, Kunst TF, et al. Phase I study of decitabine-mediated gene expression in patients with cancers involving the lungs, esophagus, or pleura. Clinical Cancer Research. 2006;**12**:5777-5785. DOI: 10.1158/1078-0432.ccr-06-0669

[116] Ansari J, Shackelford RE, El-Osta H. Epigenetics in non-small cell lung cancer: From basics to therapeutics. Translational Lung Cancer Research. 2016;**5**:155-171. DOI: 10.21037/tlcr.2016.02.02

[117] Yao H, Rahman I. Current concepts on the role of inflammation in COPD and lung cancer. Current Opinion in Pharmacology. 2009;**9**:375-383. DOI: 10.1016/j.coph.2009.06.009

[118] Shin JI, Brusselle GG. Mechanistic links between COPD and lung cancer: A role of microRNA let-7? Nature Reviews. Cancer. 2013;**14**:70. DOI: 10.1038/nrc3477-c1

Section 3

# Gene Therapy Applications

Chapter 4

# Perspective Chapter: Topoisomerase 1 and Colo Rectal Carcinoma

*Ahmed Mohamed Nabil Helaly and Doaa Ghorab*

## Abstract

Topoisomerase 1 is the main enzyme playing an important role in relaxing. The supercoiled DNA strands allow the replication fork to transcribe the DNA to RNA and finally control protein production in active and replicating cells. Blocking this essential machinery is a cornerstone mechanism in treating tumors, such as liver, breast, and metastatic colorectal carcinoma. Irinotecan is a topoisomerase inhibitor that blocks the replication ending in DNA break and tumor cell death. This chemotherapy has been successfully used in combination to overcome metastatic colorectal carcinoma. The topoisomerase-1 inhibitor makes a protein DNA complex stuck with the replicating fork creating a single DNA break, unlike topoisomerase-2, which is responsible for double DNA break. This inhibitor is exposed to drug resistance with complex machinery. Drug resistance can occur as a result of altered DNA methylation, changes in topoisomerase expression, histone recombination, or drug export pump. High expression of topoisomerase-1 is a marker of the number of tumors suggesting multiple roles of topoisomerase-1.

**Keywords:** topoisomerase-1, irinotecan, colorectal carcinoma, epigenetic, drug resistance, supercoiled DNA

## 1. Introduction

DNA topoisomerases are important enzymes that modify the double-stranded DNA topology. They act in both directions to relax supercoiled DNA and enforce relaxed DNA to be tenser. These 3D structure modifications share in the control of DNA transcription and finally protein translation. In humans, there are two types of isomerases: the first one is topoisomerase-1 (TOP-1); the second one is topoisomerase-2 (TOP-2). Because of the canonical functions of TOPs, these enzymes have been a target for treating overactive cells, such as cancer cells or bacteria that contain different types of isomerases [1]. TOPs covalently bind DNA and create DNA adducts aiming to release the over-twisted DNA strand. Topoisomerase inhibitors (TOP inhibitors) create TOP/DNA obstacles in front of the active replicating fork. The smash between the fork and the DNA protein complex ends in a DNA break. TOP-1 differs from TOP-2 in that the first one is responsible for a single DNA break, while the latter is involved in double DNA damage [2]. In this chapter, we are going to discuss the relationship between TOP-1 and cancer. The chapter aims to put a spot on the role of TOP-1 in cancer colon, and TOP-1 inhibitors' role to overcome cancer colon. It is important to discuss TOP-1 resistance.

### 1.1 Topoisomerase-1 and cancer

Since 1989, scientists recorded the high expression of TOP-1 in aggressive tumors such as late-stage colorectal carcinoma. It seems that such tumors use TOP-1 to support their replication machine. These findings were reported in renal tumors or brain neoplasms. The excess TOP-1 activity was highest seen in leukemia besides cancer colon. Moreover, TOP-1 overactivity correlated with poor prognosis in these tumors [3, 4].

TOP-1 may exert complex relation to the process of carcinogenesis. Experimental work demonstrated that knock-out TOP-1 plays a role in oncogenic property. The knock-out cells showed less aging and more replication capacity. Interestingly, the genome was more resistant to DNA damage [5]. More recent work showed opposite results concluding that oxidative stress induces cell aging with excess TOP-1 activity [6]. It is suggested that both senescence and active replication are parts of the process of neoplasia. An immunohistochemical study of secondary pterygium showed both overexpression of TOP-1 and glutathione in the same pathology [7].

A recent study on cancer liver samples using both immunohistochemistry and microarray demonstrated upregulation of TOP-1 and TOP-2 expression in poor prognosis liver cancer candidates. Both genes were involved in the function of the guardian p53 pathway and apoptosis cascade. The study showed that it is a wise idea to consider topoisomerases as oncogenes. Furthermore, these 2 targets are potentially good potentials for cancer chemotherapy [8]. TOP-1 inhibitors include etoposide, camptothecin, and Adriamycin with wide use in clinical practice. More research on different extracts to decrease the side effects and improve the efficacy is going on [9, 10]. Studies in yeast demonstrated that active or aberrant TOP-1 induces DNA mutagenicity, and later on, unstable DNA contributes to cancer development. This mechanism is controlled in mammalians by safety SUMOylation post transcription mechanism [11]. The small ubiquitin-like modifier (SUMO) is a pathway that contributes to transcription, immunity, signaling, and stabilization of the genome. Unfortunately, defective TOP-1 and its regulatory SUMOlyation may end in tumor genesis [12].

Topoisomerase-1 gene expression is considered a marker of drug response in metastatic colorectal carcinoma. It is recommended to use irinotecan (TOP-1 inhibitor) to treat these advanced colorectal tumors where TOP-1 expression is over-expressed. It seems that these tumors over-express TOP-1 to support the tumor DNA repair making the tumor cell resistant to death [13]. TOP-1 inhibitor irinotecan has been tested on cell lines to induce acetylation of the p53 and interrupt the histone deacetylase activity, making the resistant genome of growing cancer cells suitable for DNA break. It seems that TOP-1 inhibition is challenging in the understanding of the cancer colon pathway [14]. Metabolomic analysis of colorectal carcinoma cells treated by irinotecan showed shifting to glycolysis and an increase in oxygen consumption as markers of good response to cancer chemotherapy [15].

### 1.2 Topoisomerase inhibitors

The TOP-1 inhibitors were first discovered from a tree growing in China named Camptotheca acuminate. The extraction product of this tree was a component of

Chinese medicine. Later on, in the 70s and the 80s, research work managed to construct TOP-1 inhibitors in the lab and formulate them as chemotherapy [16].

Drugs inhibiting TOP-1 have been used for decades to treat malignant tumors. These chemotherapies are called TOP-1 poisons. The first-generation candidate of these compounds is camptothecin. Newer generations, include irinotecan, topotecan, and belotecan [17]. Like any chemotherapy, camptothecin derivatives have many side effects. Scientists are working with non-camptothecin generations with fewer hazards [18]. TOP-1 poisons have been prescribed for multiple tumors including colorectal cancer, ovarian tumors, small cell lung malignancy, and myeloid proliferative disorders [19]. Aggressive neoplasms over expressing TOP-1 are suggested to be good candidates for TOP-1 inhibitors as in breast or ovarian cancers besides colorectal malignancy. It is expected that TOP-1 will be radical chemotherapy for these tumors exposing their DNA to break [20–22]. The researchers were working with TOP-1 inhibitors for half a century. They managed to get the crystal structure of the enzyme. The advances in molecular docking in the last 20 years give scientists a great opportunity to design lots of derivatives to discover new drugs that are more potent with more specific functions aiming to reduce the side effects. However, the difficulty was that the mechanism of action is canonical that it is extremely difficult to avoid hazards. To overcome such obstacle, the advances in drug delivery will reduce the side effects by loading the chemotherapy dose directly onto tumor cells [23].

### 1.3 Topoisomerase-1 inhibitors and cancer colon

Irinotecan is considered the first drug of choice in treating metastatic colorectal carcinoma in combination with 5 fluorouracil and folinic acid. Irinotecan is a working TOP-1 inhibitor that is metabolized in the liver to a more active compound SN-38. This chemotherapy is characterized by a high volume of distribution. Fortunately, cancer cells have excess carboxylesterase enzymes that can metabolize irinotecan into its active component [24–26]. Irinotecan is a derivative of camptothecin and was approved for cancer colon in 1994. Its major side effects are neutropenia and diarrhea, which are responsible for dysbiosis. This chemotherapy was approved for children and adults. It is successful for metastatic colorectal neoplasia, as well as, solid tumors. The drug was constructed by the Japanese Yakult Honsha company. The newer generation products have been tested as new chemotherapy but failed in the late clinical phases such as rubitecan, gimatecan, lurtotecan, diflomotecan, elomotecan, silatecan, exatecan, namitecan [27, 28].

New compounds are being verified for TOP-1 inhibitors with better profiles. These compounds include belotecan and gimatecan. The first one is hydrophilic while the latter is dissolved in fat [29, 30].

Belotecan has been used to treat resistant ovarian tumors with better side effects in combination with other new modalities [31]. Other trials have been applied to treat lung carcinoma and biliary tumors [32, 33]. There is little data about the efficacy of belotecan on colorectal neoplasms. On the other hand, gimatecan showed promising results in brain tumors [34].

The advances in nanotechnology and drug delivery can overcome the side effects of traditional TOP-1. Nano liposomal irinotecan has been applied in a pancreatic neoplasm with the hope to achieve success. The results expressed better side effects, but the overall survival represented a weak response [35]. Nano liposomal irinotecan, in combination with other classic chemotherapy, showed better cancer pancreas response. Diarrhea seems to be less counted, but the patients still suffered from neutropenia in a fifth of cases treated to the new model [36].

In regard to colorectal cancer, there are promising clinical trials that liposomal TOP-1 will help in overcoming late-stage conditions. The experimental trials in mice showed better survival rates with fewer side effects [37]. Another strategy is to use a lipophilic active gradient of irinotecan SN38 to suppress advanced colorectal tumors. The results showed a successful modality and the FDA approved the drug for the treatment of late-stage cancer colon [38].

New drugs, which are non-camptothecin derivatives, have been introduced to manage different tumors. These modalities include dibenzonaphthyridines, as well as, indeno isoquinolines. They are more strong stable DNA-protein complexes. Trials of metals such as platinum, gold, copper, zinc, and others have been suggested as TOP-1 modulators [39].

More research work is concerned with indenoisoquinolines, regards the capability to inhibit TOP-1. These new groups of the drug are subjected to structural modification. They represent promising drugs with potentially fewer side effects. It is proposed that the new chemical will have a multi-mechanism of action like nuclear receptors targeting, TOP-2 interaction, modulating estrogen receptors, and manipulating VEGFR and HIF-1 alpha [40]. Studies suggested that indenoisoquinolines are weak TOP-1 inhibitors in combination with other functions. It seems that weak TOP-1 inhibitors in conjunction with other drugs, or even other pathways, are good rationales for modern cancer therapy [41].

### *1.3.1 Indenoisoquinolines*

More than 20 years ago, a group of scientists managed to create a new anti-TOP-1 chemical 6,11-dimethyl-6,11-dihydro-5H-indeno[1,2-c]isoquinolin-5-one. This compounds the parent of quinolone family, and it exerted promising results on different human cell lines [42]. Since that time, dozens of derivatives have been introduced to be tested experimentally. These compounds have the advantage of being flat ones that strongly bind the TOP-1/DNA hybrid. The new compounds bind the TOP-1 at the arginine 364 of the enzyme TOP-1 [43]. These potential new drugs showed stronger cytotoxicity relative to their anti-TOP-1 mechanism and are expected with combined dynamic to have a less toxic profile [44]. Other pathways affected by indenoisoquinolines include induction of cell cycle arrest, stimulation of apoptotic response, and modulation of MAPK cascade. Other potential role includes phosphorylation of JUNK pathway and inhibition of the oncogene MYC pathway [45]. On the other hand, despite the strong multi-dynamic function of these drugs, they did not replace the classic TOP-1. Research is still going to tune drugs with a balance between efficacy and hazard effects.

## 1.4 Topoisomerase-1 inhibitor resistance

Topoisomerase inhibitors, like other chemotherapy, are subjected to resistance. Many mechanisms are involved in this process. The tumor uses efflux mechanisms to reduce the level of the drug in the tumor mass. Mutations of the TOP-1 make the drug less effective in inhibiting the enzyme. Other strategies include enhancing the DNA repair to overcome TOP-1 poison. The tumor cells stimulate p53 to support the malignant mass to survive and suppress apoptosis. Limiting the drug's bioavailability can be another way to overcome chemotherapy [46].

Recent work put the spot on cancer colon stem cell role in the development of cancer. These highly replicating cells can overexpress the ATP cassette transporters.

The experimental work showed that the over-expressed MYC oncogene supports the tumor resistance by enriching ATP transporters [47]. The ATP cassette sub-family G isoform 2 is expected to overcome xenobiotic effects with abundance in the GIT and near blood vessels. Besides their role in cancer stem cells, they represent a defense mechanism against chemotherapy. Tyrosine kinase inhibitors, phosphodiesterase-5 inhibitors, and the fumitremorgin-type indolyl diketopiperazine have been used to support chemotherapy to overcome resistance by inhibiting the ATP transporters [48]. The resistance was marked in cases associated with the marker ATP cassette type G group member 2 [49]. To overcome the resistance obstacles, the use of an ATP cassette inhibitor has been introduced. Another method was to apply the new TOP-1 inhibitor FL118, which possesses the capacity to overcome efflux resistance. This new analog is weakly transported by ATP carriers [50]. KO143 is a potent antagonist of ATP cassette sub-family G member 2. It is stable and not easily metabolized by the cytochrome enzymes in the liver [51].

ATP-binding transporters have been involved in the resistance to TOP-1 inhibitors in cancer breast. Experimental work on cancer colon cell lines concluded that colon tumor cells expressed a similar mechanism. Furthermore, it is possible to add compounds inhibiting efflux mechanisms such as SCO-201 to the chemotherapy regimen to sensitize the tumor to respond to the chemotherapy indicating better survival rates [52]. It is important to notice that overexpression of the ATP transporters is a marker of resistance to cancer colon. It is recorded that candidates' higher expression is most likely to resist camptothecin derivatives.

#### *1.4.1 ATP efflux mechanism*

The master mechanism of resistance of colorectal tumors to chemotherapy is drug efflux mechanism. The ATP-binding cassette sub-family G member 2 (ABCG2) is responsible for decreasing anti-TOP-1 bioavailability inside the tumor microenvironment leading to failure of the therapy [53]. The promoter regulating the expression of these shuttles is epigenetically regulated by methylation. Aggressive tumors express high ATP cassettes to get rid of the xenobiotic load. Hypomethylation of the promotor-regulating ATP transporter was recorded in different tumor cell lines [54]. The ABCG2 is formed from a sequence of 655 amino acids weighing 72-kilo Dalton. The structure is homodimer with two nucleotide-binding sites to export the drugs [55, 56]. The transporters play important role in the maintenance of the tumor microenvironment. They keep the balance of osmotic pressure. They have another role in antigen processing and modulation of cell division. Cell trafficking and cholesterol metabolism are affected by the ATP transporters [57].

Recently, $N^6$-methyladenosine ($m^6A$) RNA modification has been a new modality to treat ATP transporter through epigenetic mechanism at the RNA level [58].

Another important mechanism to defeat chemotherapy is the mutation of DNA repair response. In cancer cells, positive mutations make the tumor resistant to TOP-1 effects. The tumor genes such as BRCA1, BRCA2, PLAB2, and BARD1 have been mutated. Studies recorded alterations in 86 genes related to DNA repair. Such overactivity makes the single DNA break to the TOP-1 less effective [59].

Cancer colon cells expressing both SENP-1 (sentrin specific protease) and HIF-1α (Hypoxia inducing factor alpha) are resistant to chemotherapy. Both of them overexpress proteins SUMO pathway and make the tumor resistant to hypoxia [60]. Recently, SENP1 is a target of new compounds to develop a new regimen to overcome resistance [61].

Cancer cells can develop an alternative way to resist chemotherapy, specially irinotecan by augmenting the detoxification pathway. The active metabolite SN-38 is inactivated by glucuronidation making the drug short of killing the cancer cells. The nuclear receptors pregnane X receptors and steroid/xenobiotic receptors enhance the metabolism process of SN-38. It is important to notice that they are expressed in excess in the liver and the GIT. They also induce a battery of genes that express xenobiotic transporters that make irinotecan ineffective within a short period [62].

To repel TOP-1 inhibitors, the tumor cells induce a high copy number of the TOP-1 gene to make the drug subtherapeutic to kill the neoplastic cells. Furthermore, the TOP-1 is subjected to chromosomal alterations to be transcribed from different loci. These overexpressed loci are considered biomarkers of the drug response later on [13, 63].

The cancer colon cells are smart enough to create mutated TOP-1 that is not responding to the classic inhibitors. Experimental studies on resistant cell lines demonstrated several mutations that resist camptothecin derivatives or the new generation TOP-1 drugs [64].

The antioxidant balance in the cancer metabolism plays important role in protecting the tumor cells from xenobiotic toxicity and oxidative stress. One of the common mechanisms to support the redox is to express a high amount of glutathione reductase allowing the tumor to grow in unnatural situations and protect the mass from chemotherapy [65].

Another model for TOP-1 inhibitor resistance is colorectal tumors expressing active epidermal growth factor (EGFR). It was recorded that the active metabolite SN-38 is subjected to resistance because of the over-expression of EGFR. This factor triggers a trophic response in cancer growth via a cascade of a signaling pathway [66, 67].

### 1.5 Epigenetic therapy and cancer colon

New work has applied epigenetic modifier agents in combination with classic chemotherapy with a promising response. The epigenetic changes improved the drug response and reduce the needed dose to improve the clinical outcome. These transformers include DNA methyl transferase, decitabine, azacytidine, and zebularine. The study also used drugs that were considered histone deacetylase inhibitors including the well-known mood stabilizer valproic acid with promising results [68]. During the process of cancer evolution, cancer cells have different strategies to survive. It was recorded that p53 is mutated and histone deacetylases (HDACs) are over-expressed. It was considered that this epigenetic mechanism corresponds to drug resistance. Experimental inhibition of HDACs by small hairpin RNAs resulted in a better response of chemotherapy against the resistant SW480 cell lines. Indeed, epigenetic modification is part of the process of carcinogenesis and drug resistance response [69].

The DNA topology is inherited by an epigenetic mechanism. However, the process is unclear. The status of DNA twisting is a unique criterion of each cell type. This information is considered a cellular memory that is transmitted from the parent cells to the daughters by mitosis. The status of DNA 3D structure controls transcription and gene expression [70]. It has been recorded that parent cells with excess supercoiling deliver daughter cells with the impaired cell cycle. DNA supercoiling is important in the process of chromatin condensation. Previously, it was thought that DNA remains quiet during the process of mitosis. However, recent evidence that a battery of genes is still active in the mitosis process to be involved in the development of the next generation of cells [71]. The activity of the TOP-1 allows the chromatin to adapt to the DNA-positive supercoils. On the other hand, an excess of negative supercoils ends in DNA/RNA hybrid structure with resulting in DNA damage [72]. Although, DNA/RNA hybrid is considered a mechanism of regulating gene expression called an R loop [73].

### 1.6 Topoisomerase-1 biomarkers of response

The first line of treatment for wild-type RAS metastatic colon cancer was the anti-EGFR immunotherapy cetuximab or panitumumab [74]. Experimentation searched for the colorectal cancer biomarkers in the blood. Recently, kits to extract the tumor cDNA are available clinically to evaluate associated parameters predicting the response of the tumor to classic chemotherapy. Moreover, liquid samples assaying RAS and BRAF oncogenes have a role in chemotherapy of colorectal cancer. The assays showed that tumors with wild-type RAS and BRAF were more responsive to camptothecin chemotherapy in combination with immunotherapy as a third line of treatment for colorectal metastasis [75]. The selection of the biomarkers as a strategy of personalized medicine in colon tumors reduces the side effects for precisely predicted response to the combined TOP-1 chemotherapy-related regimen. These markers include *DYPD*, *UGT1A1, HPP1*, *HER2*, *HER3, PIK3CA*, and *PTEN* [76]. It is important to say that EGFR is modulating a battery-controlling factor or biomarker that can predict the tumor response. It influences RAS/BRAF/MEK/MAPK and PI3K/PTEN/AKT cascades [77].

### 1.7 Topoisomerase and other diseases

TOP-1 is an essential enzyme in the machinery of DNA transcription by modifying the topology of the DNA helix, and it is an important factor in stabilizing the genome. Furthermore, this enzyme has another role in the process of transcription in a way not directly related to unwinding DNA. Researchers proposed TOP-1 targets as a potential treatment for autism [78]. Mitochondria depend on the imported TOP-1 to relax the DNA during the process of transcription. Mitochondrial DNA damage is involved in many, such as neurodegenerative disorders and cancer. TOP-1 with its double role is a valuable object of cancer chemotherapy [79].

Another important scope of drugs inhibiting TOP-1 is modulating the immune response on exposure to microorganisms. Experimental studies showed that loading mice with endotoxins expressed a less harmful immune reaction than candidates tested with TOP-1 inhibitors. Low-dose TOP-1 chemotherapy such as camptothecin improved the antiviral unfavorable reaction. Interestingly, the low-dose poison with mild DNA break improved the situation in fighting the viral load [80].

Recently during the COVID-19 crisis, topotecan (TOP-1) has been applied to reduce the aberrant immune response to corona infection. Experimental work on hamsters and mice showed that a topotecan dosing improved the immune response. The TOP-1 inhibitors may exert antiviral capacity [81].

#### *1.7.1 Mutated TOP-1*

Mutated TOP-1 was a part of different chronic disorders. Malfunctioning TOP-1 is associated with unstable DNA. The causes of such complex pathology are still unknown. The hybrid TOP-1/DNA abnormal complex works as a TOP-1 inhibitor ending in excess DNA break. These phenomena were recorded in massive spinocerebellar disorders. It was reported that ATM (ataxia telangiectasia, mutated), a serine/threonine protein kinase, plays an essential role in cell cycle activation, chromatin remodeling, DNA repair, or stimulation of apoptosis. Experimental knockout of ATM showed accumulated pathogenic TOP-1/DNA complex with excess DNA break creating neurodegenerative pathology in animal models. Furthermore, healthy ATM can mediate TOP-1 complex ubiquitination [82]. These results established the idea that healthy TOP-1 is essential for development of central nervous system.

### 1.8 Innovations and novel strategies from a clinical perspective

Recently, the advances in multi-omic studies provided a signature of metastatic colorectal carcinoma cases aiming to predict the response of the chemotherapy. Metastatic candidates expressed unique proteomic profiles. There is a distinct profile for colorectal tumors, associated with liver metastasis. As regarding the genomic profile, there was no significant difference between early or late-stage tumors. These findings build up personalized strategies to treat or predict poor prognosis. Metastatic colorectal patients showed overexpression of oxidative phosphorylation and Krebs cycle enzymes [58]. A new strategy to use drugs that inhibit the rate of metabolism has been introduced in clinical trials as an adjuvant in colorectal metastasis in the liver. This modality is considered as tumor micro immunity reprogramming [83].

On the other hand, the analysis of circulating tumor DNA in the blood showed a mutation signature related to the tumor outcome. The status of RAS/BRAF mutations correlated with the prognosis of patients. The RAS/BRAF mutation load correlated with overall survival. As fewer mutations were detected in the blood sample, higher rates of remission were detected [84]. Another clinical study showed a similar profile. Excessive mutation RAS/BRAF suggested poor response to the first line of treatment of colorectal carcinoma [85].

A new modality mXELIRI (capecitabine plus irinotecan) has been recently applied with or without immunotherapy. The patients tolerated therapy for colorectal tumors with acceptable efficacy and toxicity profile [86]. Other trials have been applied to combine irinotecan and vincristine in soft tissue tumors [87].

New combination chemotherapy has been introduced recently. The anti-HER3 antibody patritumab has been added to the new TOP-1 inhibitor DX-8951 derivative (DXd) to treat resistant colorectal cancer with potential success [88].

## 2. Conclusions

TOP-1 gene and TOP-1 target protein are important in both carcinogenesis and pharmacology of different tumors. It is concluded that TOP-1 classic inhibitors **Table 1** are the main line of treatment of metastatic colorectal carcinoma. These drugs with complex mechanism of action are associated with profound side effects and subjected to different types of machinery of drug resistance. The discovery of new drugs with multi dynamics, including mild TOP-1 in conjunction with other actions, is a promising modality to bypass the toxic effects of classic TOP-1 inhibitors as shown in **Table 2**. Too strong anti-TOP is no longer a wise strategy to apply because of the canonical TOP function in almost every cell. Epigenetic tools can be added to improve the chemotherapy outcome.

| Drug | The study | Cell line | Reference |
|---|---|---|---|
| Hydroxycampthecin | The drug inhibit TGF-beta1 and inhibit fibrosis | Human fibroblast | [89] |
| Homocamptothecin E-beta ring hydroxylactone | New anti TOP1 | Molecular docing | [90] |
| 1,3-disubstituted-4-hydroxy-6-methylpyridin-2(1H)-one | New anti TOP1 Managed to treat leishmania with good cytoxicity profile | Leishmania | [91] |
| 7,12-dihydrodibenzo[b,h][1,6] naphthyridine and 7H-Chromeno[3,2-c] quinoline derivatives | Non campthecin TOP1 inhibitor | Human cancer cell lines (A549 and MCF-7) and in silico | [92] |
| 9-Aminocamptothecin (9-AC) | Anti TOP1 for prostate cancer with promising results | Prostate cell lines | [93] |
| Metallated porphyrins | Ant TOP1 for different cell lines | K562, U937, HL-60, Jurkat, A549 and HeLa cancer cell lines | [94] |
| Heteroliptic cupper | Anti TOP for mulriple cancer cell lines with success | Human lung (A549), cervical (HeLa) and colon (HCT-15) | [95] |
| Irintotecan | Alter metabolism of gut | Enterocytes | [96] |

**Table 1.**
*In vitro studies with TOP1 inhibitors.*

| The drug | The study | The host | Reference |
|---|---|---|---|
| Exatecan derivative (DX-8951 derivative, DXd) | The drug was combined with the immunotherapy DS-8201a is a human epidermal growth factor receptor 2 (HER2) | HER2-positive NCI-N87 cells and HER2-negative MDA-MB-468-Luc cells incubated in mice | [97] |
| Topotecan (TPT) | Potential role in COVID patients with success in reducing the immune storm | Transgenic mice | [81] |
| Marine alkaloid lamellarin D drivatives | More effective in treating cancer colon | Murine colon cancer | [98] |
| AZD2014 or INK128 in combination with irinotecan | Induce apoptosis in Malignant peripheral nerve sheath tumors | Zebra fish | [99] |
| 7-Ethyl-10-hydroxy-camptothecin (SN38), the active metabolite of irinotecan (CPT-11) | Liver breast and colon cancer model expressed better response | Both in vivo and in vitro | [100] |
| Quercetin | A flavonoid with anti TOP anti-gastric cancer | Both in vivo/in vitro | [101] |

**Table 2.**
*In vivo studies with TOP1 inhibitor.*

## Author details

Ahmed Mohamed Nabil Helaly[1,2*] and Doaa Ghorab[3,4]

1 Faculty of Medicine, Forensic Medicine, and Clinical Toxicology Department, Mansoura University, Egypt

2 Faculty of Medicine, Clinical Science Department, Yarmouk University, Irbid, Jordan

3 Faculty of Medicine, Pathology Department, Mansoura University, Egypt

4 Faculty of Medicine, Basic Science, Yarmouk University, Irbid, Jordan

*Address all correspondence to: ahmedhelaly@mans.edu.eg

## References

[1] Bush NG, Evans-Roberts K, Maxwell A. DNA topoisomerases. EcoSal Plus. 2015;**6**:2

[2] Nitiss JL, Kiianitsa K, Sun Y, Nitiss KC, Maizels N. Topoisomerase assays. Current Protocol. 2021;**1**(10):e250

[3] Husain I, Mohler JL, Seigler HF, Besterman JM. Elevation of topoisomerase I messenger RNA, protein, and catalytic activity in human tumors: Demonstration of tumor-type specificity and implications for cancer chemotherapy. Cancer Research. 1994;**54**(2):539-546

[4] Zhao C, Yasui K, Lee CJ, Kurioka H, Hosokawa Y, Oka T, et al. Elevated expression levels of NCOA3, TOP1, and TFAP2C in breast tumors as predictors of poor prognosis. Cancer. 2003;**98**(1):18-23

[5] Humbert N, Martien S, Augert A, Da Costa M, Mauen S, Abbadie C, et al. A genetic screen identifies topoisomerase 1 as a regulator of senescence. Cancer Research. 2009;**69**(10):4101-4106

[6] Bu H, Baraldo G, Lepperdinger G, Jansen-Dürr P. mir-24 activity propagates stress-induced senescence by down regulating DNA topoisomerase 1. Experimental Gerontology. 2016;**75**:48-52

[7] Doaa G, Helaly AM. Badawi a dual pathogenesis of primary and recurrent pterygium: Immunohistochemical proof. The Open Ophthalmology. 2021;**15**:229-235. DOI: 10.2174/1874364102115010229

[8] Liu LM, Xiong DD, Lin P, Yang H, Dang YW, Chen G. DNA topoisomerase 1 and 2A function as oncogenes in liver cancer and may be direct targets of nitidine chloride. International Journal of Oncology. 2018;**53**(5):1897-1912

[9] Kathiravan MK, Kale AN, Nilewar S. Discovery and development of topoisomerase inhibitors as anticancer agents. Mini Reviews in Medicinal Chemistry. 2016;**16**(15):1219-1229

[10] Nukuzuma S, Nakamichi K, Kameoka M, Sugiura S, Nukuzuma C, Tasaki T, et al. Suppressive effect of topoisomerase inhibitors on JC polyomavirus propagation in human neuroblastoma cells. Microbiology and Immunology. 2016;**60**(4):253-260

[11] Li M, Pokharel S, Wang JT, Xu X, Liu Y. RECQ5-dependent SUMOylation of DNA topoisomerase I prevents transcription-associated genome instability. Nature Communications. 2015;**6**:6720

[12] Han ZJ, Feng YH, Gu BH, Li YM, Chen H. The post-translational modification, SUMOylation, and cancer (Review). International Journal of Oncology. 2018;**52**(4):1081-1094

[13] Braun MS, Richman SD, Quirke P, Daly C, Adlard JW, Elliott F, et al. Predictive biomarkers of chemotherapy efficacy in colorectal cancer: Results from the UK MRC FOCUS trial. Journal of Clinical Oncology. 2008a;**26**(16):2690-2698

[14] Marx C, Sonnemann J, Beyer M, Maddocks ODK, Lilla S, Hauzenberger I, et al. Mechanistic insights into p53-regulated cytotoxicity of combined entinostat and irinotecan against colorectal cancer cells. Molecular in Oncology. 2021;**15**(12):3404-3429

[15] MarxC,SonnemannJ,MaddocksODK, Marx-Blümel L, Beyer M, Hoelzer D, et al. Global metabolic alterations in colorectal cancer cells during

irinotecan-induced DNA replication stress. Cancer Metabolism. 2022;**10**(1):10

[16] Wall ME, Wani MC. Camptothecin and taxol: Discovery to clinic--thirteenth Bruce F. Cancer Research. 1995;**55**(4):753-760

[17] Martín-Encinas E, Selas A, Palacios F, Alonso C. The design and discovery of topoisomerase I inhibitors as anticancer therapies. Expert Opinion in Drug Discovery. 2022;**17**(6):581-601

[18] Talukdar A, Kundu B, Sarkar D, Goon S, Mondal MA. Topoisomerase I inhibitors: Challenges, progress and the road ahead. European Journal of Medicinal Chemistry. 2022;**236**:114304

[19] Rasheed ZA, Rubin EH. Mechanisms of resistance to topoisomerase I-targeting drugs. Oncogene. 2003;**22**(47):7296-7304

[20] O'Connor MJ. Targeting the DNA damage response in Cancer. Molecular Cell. 2015;**60**(4):547-560

[21] Pommier Y. Topoisomerase I inhibitors: Camptothecins and beyond. Nature Reviews. Cancer. 2006;**6**(10):789-802

[22] Xu Y, Her C. Inhibition of topoisomerase (DNA) I (TOP1): DNA damage repair and anticancer therapy. Biomolecules. 2015;**5**(3):1652-1670

[23] Cinelli MA. Topoisomerase 1B poisons: Over a half-century of drug leads, clinical candidates, and serendipitous discoveries. Medicinal Research Reviews. 2019;**39**(4):1294-1337

[24] Guichard S, Terret C, Hennebelle I, Lochon I, Chevreau P, Frétigny E, et al. CPT-11 converting carboxylesterase and topoisomerase activities in tumour and normal colon and liver tissues. British Journal of Cancer. 1999;**80**(3-4):364-370

[25] Mathijssen RH, van Alphen RJ, Verweij J, Loos WJ, Nooter K, Stoter G, et al. Clinical pharmacokinetics and metabolism of irinotecan (CPT-11). Clinical Cancer Research. 2001;**7**(8):2182-2194

[26] Slatter JG, Schaaf LJ, Sams JP, Feenstra KL, Johnson MG, Bombardt PA, et al. Pharmacokinetics, metabolism, and excretion of irinotecan (CPT-11) following I.V. infusion of [(14)C]CPT-11 in cancer patients. Drug Metabolism and Disposition. 2000;**28**(4):423-433

[27] Bailly C. Irinotecan: 25 years of cancer treatment. Pharmacological Research. 2019;**148**:104398

[28] Li F, Jiang T, Li Q, Ling X. Camptothecin (CPT) and its derivatives are known to target topoisomerase I (Top1) as their mechanism of action: Did we miss something in CPT analogue molecular targets for treating human disease such as cancer? American Journal of Cancer Research. 2017;**7**(12):2350-2394

[29] Covey JM, Jaxel C, Kohn KW, Pommier Y. Protein-linked DNA strand breaks induced in mammalian cells by camptothecin, an inhibitor of topoisomerase I. Cancer Research. 1989;**49**(18):5016-5022

[30] Thomas A, Pommier Y. Targeting topoisomerase I in the era of precision medicine. Clinical Cancer Research. 2019;**25**(22):6581-6589

[31] Hur J, Ghosh M, Kim TH, Park N, Pandey K, Cho YB, et al. Synergism of AZD6738, an ATR inhibitor, in combination with belotecan, a Camptothecin analogue, in chemotherapy-resistant ovarian Cancer. International Journal of Molecular Sciences. 2021;**22**(3):1223

[32] Jun E, Park Y, Lee W, Kwon J, Lee S, Kim MB, et al. The identification of candidate effective combination regimens for pancreatic cancer using the histoculture drug response assay. Scientific Reports. 2020;**10**(1):12004

[33] Kang JH, Lee KH, Kim DW, Kim SW, Kim HR, Kim JH, et al. A randomised phase 2b study comparing the efficacy and safety of belotecan vs. topotecan as monotherapy for sensitive-relapsed small-cell lung cancer. British Journal of Cancer. 2021;**124**(4):713-720

[34] Cesare DE. High efficacy of intravenous Gimatecan on human tumor xenografts. Anticancer Research. 2018;**38**(10):5783-5790

[35] Kasi A, McGinnis T, Naik G, Handa S, Williams G, Paluri R. Efficacy and tolerability of the combination of nano-liposomal irinotecan and 5-fluorouracil/leucovorin in advanced pancreatic adenocarcinoma: Post-approval clinic experience. Journal of Gastrointestinal Oncology. 2021;**12**(2):464-473

[36] Yasuoka H, Naganuma A, Kurihara E, Kobatake T, Ijima M, Tamura Y, et al. Efficacy and safety of the combination of Nano-liposomal irinotecan and 5-fluorouracil/L-leucovorin in unresectable advanced pancreatic cancer: A real-world study. Oncology. 2022;**100**(8):449-459

[37] Liu X, Jiang J, Chan R, Ji Y, Lu J, Liao YP, et al. Improved efficacy and reduced toxicity using a custom-designed irinotecan-delivering silicasome for orthotopic colon cancer. ACS Nano. 2019;**13**(1):38-53

[38] Xing J, Zhang X, Wang Z, Zhang H, Chen P, Zhou G, et al. Novel lipophilic SN38 prodrug forming stable liposomes for colorectal carcinoma therapy. International Journal of Nanomedicine. 2019;**14**:5201-5213

[39] Gokduman K. Strategies targeting DNA topoisomerase I in Cancer chemotherapy: Camptothecins, nanocarriers for camptothecins, organic non-Camptothecin compounds and metal complexes. Current Drug Targets. 2016;**17**(16):1928-1939

[40] Cushman M. Design and synthesis of Indenoisoquinolines targeting topoisomerase I and other biological macromolecules for Cancer chemotherapy. Journal of Medicinal Chemistry. 2021;**64**(24):17572-17600

[41] Wang KB, Elsayed MSA, Wu G, Deng N, Cushman M, Yang D. Indenoisoquinoline topoisomerase inhibitors strongly bind and stabilize the *MYC* promoter G-Quadruplex and downregulate *MYC*. Journal of American Chemical Society. 2019;**141**(28):11059-11070

[42] Cho WJ, Park MJ, Imanishi T, Chung BH. A novel synthesis of benzo[c] phenanthridine skeleton and biological evaluation of isoquinoline derivatives. Chemical Pharmacy Bulletin (Tokyo). 1999;**47**(6):900-902

[43] Khadka DB, Le QM, Yang SH, Van HT, Le TN, Cho SH, et al. Design, synthesis and docking study of 5-amino substituted indeno[1,2-c]isoquinolines as novel topoisomerase I inhibitors. Bioorganic & Medicinal Chemistry. 2011;**19**(6):1924-1929

[44] Strumberg D, Pommier Y, Paull K, Jayaraman M, Nagafuji P, Cushman M. Synthesis of cytotoxic indenoisoquinoline topoisomerase I poisons. Journal of Medicinal Chemistry. 1999;**42**(3):446-457

[45] Park EJ, Kiselev E, Conda-Sheridan M, Cushman M, Pezzuto JM. Induction

of apoptosis by 3-amino-6-(3-aminopropyl)-5,6-dihydro-5,11-dioxo-11H-indeno[1,2-c]isoquinoline via modulation of MAPKs (p38 and c-Jun N-terminal kinase) and c-Myc in HL-60 human leukemia cells. Journal of Natural Products. 2012;**75**(3):378-384

[46] Tomicic MT, Kaina B. Topoisomerase degradation, DSB repair, p53 and IAPs in cancer cell resistance to camptothecin-like topoisomerase I inhibitors. Biochimica et Biophysica Acta. 2013;**1835**(1):11-27

[47] Zhang HL, Wang P, Lu MZ, Zhang SD, Zheng L. c-Myc maintains the self-renewal and chemoresistance properties of colon cancer stem cells. Oncology Letters. 2019;**17**(5):4487-4493

[48] Stacy AE, Jansson PJ, Richardson DR. Molecular pharmacology of ABCG2 and its role in chemoresistance. Molecular Pharmacology. 2013;**84**(5):655-669. DOI: 10.1124/mol.113.088609

[49] Tuy HD, Shiomi H, Mukaisho KI, Naka S, Shimizu T, Sonoda H, et al. ABCG2 expression in colorectal adenocarcinomas may predict resistance to irinotecan. Biochimica et Biophysica Acta. 2016;**12**(4):2752-2760

[50] Westover D, Ling X, Lam H, Welch J, Jin C, Gongora C, et al. FL118, a novel camptothecin derivative, is insensitive to ABCG2 expression and shows improved efficacy in comparison with irinotecan in colon and lung cancer models with ABCG2-induced resistance. Molecular Cancer. 2015;**14**:92

[51] Liu K, Zhu J, Huang Y, Li C, Lu J, Sachar M, et al. Metabolism of KO143, an ABCG2 inhibitor. Drug Metabolism and Pharmacokinetics. 2017;**32**(4):193-200

[52] Ambjørner SEB, Wiese M, Köhler SC, Svindt J, Lund XL, Gajhede M, et al. The pyrazolo[3,4-d]pyrimidine derivative, SCO-201, reverses multidrug resistance mediated by ABCG2/BCRP. Cell. 2020;**9**(3):613

[53] Moon HH, Kim SH, Ku JL. Correlation between the promoter methylation status of ATP-binding cassette sub-family G member 2 and drug sensitivity in colorectal cancer cell lines. Oncology Reports. 2016;**35**(1):298-306

[54] Spitzwieser M, Pirker C, Koblmüller B, Pfeiler G, Hacker S, Berger W, et al. Promoter methylation patterns of ABCB1, ABCC1 and ABCG2 in human cancer cell lines, multidrug-resistant cell models and tumor, tumor-adjacent and tumor-distant tissues from breast cancer patients. Oncotarget. 2016;7(45):73347-73369

[55] Litman T, Brangi M, Hudson E, Fetsch P, Abati A, Ross DD, et al. The multidrug-resistant phenotype associated with overexpression of the new ABC half-transporter, MXR (ABCG2). Journal of Cell Science. 2000;**113**(Pt 11):2011-2021

[56] Xu J, Liu Y, Yang Y, Bates S, Zhang JT. Characterization of oligomeric human half-ABC transporter ATP-binding cassette G2. The Journal of Biological Chemistry. 2004;**279**(19):19781-19789

[57] Liu X. ABC family transporters. Advances in Experimental Medicine and Biology. 2019;**1141**:13-100

[58] Li C, Sun YD, Yu GY, Cui JR, Lou Z, Zhang H, et al. Integrated omics of metastatic colorectal Cancer. Cancer Cell. 2020;**38**(5):734-747

[59] Cuella-Martin R, Hayward SB, Fan X, Chen X, Huang JW, Taglialatela A, et al. Functional interrogation of DNA damage response variants with base editing screens. Cell. 2021;**184**(4):1081-1097

[60] Chen MC, Nhan DC, Hsu CH, Wang TF, Li CC, Ho TJ, et al. SENP1 participates in irinotecan resistance in human colon cancer cells. Journal of Cellular Biochemistry. 2021;**122**(10):1277-1294

[61] Wei J, Wang H, Zheng Q, Zhang J, Chen Z, Wang J, et al. Recent research and development of inhibitors targeting sentrin-specific protease 1 for the treatment of cancers. European Journal of Medicinal Chemistry. 2022;**241**:114650

[62] Ozawa S, Miura T, Terashima J, Habano W. Cellular irinotecan resistance in colorectal cancer and overcoming irinotecan refractoriness through various combination trials including DNA methyltransferase inhibitors: A review. Cancer Drug Resistant. 2021;**4**(4):946-964

[63] Rømer MU, Jensen NF, Nielsen SL, Müller S, Nielsen KV, Nielsen HJ, et al. TOP1 gene copy numbers in colorectal cancer samples and cell lines and their association to in vitro drug sensitivity. Scandinavian Journal of Gastroenterology. 2012;**47**(1):68-79

[64] Jensen NF, Agama K, Roy A, Smith DH, Pfister TD, Rømer MU, et al. Characterization of DNA topoisomerase I in three SN-38 resistant human colon cancer cell lines reveals a new pair of resistance-associated mutations. Journal of Experimental & Clinical Cancer Research. 2016;**35**:56

[65] Narayanankutty A, Job JT, Narayanankutty V. Glutathione, an antioxidant tripeptide: Dual roles in carcinogenesis and chemoprevention. Current Protein & Peptide Science. 2019;**20**(9):907-917

[66] Asano T. Drug resistance in cancer therapy and the role of epigenetics. Journal of Nippon Medical School. 2020;**87**(5):244-251

[67] Petitprez A, Larsen AK. Irinotecan resistance is accompanied by upregulation of EGFR and Src signaling in human cancer models. Current Pharmaceutical Design. 2013;**19**(5):958-964

[68] Hosokawa M, Tanaka S, Ueda K, Iwakawa S. Different schedule-dependent effects of epigenetic modifiers on cytotoxicity by anticancer drugs in colorectal Cancer cells. Biological & Pharmaceutical Bulletin. 2017;**40**(12):2199-2204

[69] Alzoubi S, Brody L, Rahman S, Mahul-Mellier AL, Mercado N, Ito K, et al. Synergy between histone deacetylase inhibitors and DNA-damaging agents is mediated by histone deacetylase 2 in colorectal cancer. Oncotarget. 2016;**7**(28):44505-44521

[70] Jha RK, Levens D, Kouzine F. Mechanical determinants of chromatin topology and gene expression. Nucleus. 2022;**13**(1):94-115

[71] Palozola KC, Donahue G, Liu H, Grant GR, Becker JS, Cote A, et al. Mitotic transcription and waves of gene reactivation during mitotic exit. Science. 2017;**358**(6359):119-122

[72] Edwards DS, Maganti R, Tanksley JP, Luo J, Park JJH, Balkanska-Sinclair E, et al. BRD4 prevents R-loop formation and transcription-replication conflicts by ensuring efficient transcription elongation. Cell Reports. 2020;**32**(12):108166

[73] Yuan W, Al-Hadid Q, Wang Z, Shen L, Cho H, Wu X, et al. TDRD3 promotes DHX9 chromatin recruitment and R-loop resolution. Nucleic Acids Research. 2021;**49**(15):8573-8591

[74] Van Cutsem E, Cervantes A, Adam R, Sobrero A, Van

Krieken JH, Aderka D, et al. ESMO consensus guidelines for the management of patients with metastatic colorectal cancer. Annals of Oncology. 2016;**27**(8):1386-1422

[75] Cremolini C, Rossini D, Dell'Aquila E, Lonardi S, Conca E, Del Re M, et al. Rechallenge for patients with RAS and BRAF wild-type metastatic colorectal cancer with acquired resistance to first-line cetuximab and irinotecan: A phase 2 single-arm clinical trial. JAMA Oncology. 2019;**5**(3):343-350

[76] Taieb J, Jung A, Sartore-Bianchi A, Peeters M, Seligmann J, Zaanan A, et al. The evolving biomarker landscape for treatment selection in metastatic colorectal Cancer. Drugs. 2019;**79**(13):1375-1394

[77] Peeters M, Price TJ, Cervantes A, Sobrero AF, Ducreux M, Hotko Y, et al. Randomized phase III study of panitumumab with fluorouracil, leucovorin, and irinotecan (FOLFIRI) compared with FOLFIRI alone as second-line treatment in patients with metastatic colorectal cancer. Journal of Clinical Oncology. 2010;**28**(31):4706-4713

[78] Li M, Liu Y. Topoisomerase I in human disease pathogenesis and treatments. Genomics Proteomics Bioinformatics. 2016;**14**(3):166-171

[79] Das BB, Ghosh A, Bhattacharjee S, Bhattacharyya A. Trapped topoisomerase-DNA covalent complexes in the mitochondria and their role in human diseases. Mitochondrion. 2021;**60**:234-244

[80] Pépin G, Nejad C, Ferrand J, Thomas BJ, Stunden HJ, Sanij E, et al. Topoisomerase 1 inhibition promotes cyclic GMP-AMP synthase-dependent antiviral responses. MBio. 2017;**8**(5):e01611-e01617

[81] Ho JSY, Mok BW, Campisi L, Jordan T, Yildiz S, Parameswaran S, et al. TOP1 inhibition therapy protects against SARS-CoV-2-induced lethal inflammation. Cell. 2021;**184**(10):2618-2632

[82] Katyal S, Lee Y, Nitiss KC, Downing SM, Li Y, Shimada M, et al. Aberrant topoisomerase-1 DNA lesions are pathogenic in neurodegenerative genome instability syndromes. Nature Neuroscience. 2014;**17**(6):813-821

[83] Wu Y, Yang S, Ma J, Chen Z, Song G, Rao D, et al. Spatiotemporal immune landscape of colorectal Cancer liver metastasis at single-cell level. Cancer Discovery. 2022;**12**(1):134-153

[84] Wang F, Huang YS, Wu HX, Wang ZX, Jin Y, Yao YC, et al. Genomic temporal heterogeneity of circulating tumour DNA in unresectable metastatic colorectal cancer under first-line treatment. Gut. 2022;**71**(7):1340-1349

[85] Yao J, Zang W, Ge Y, Weygant N, Yu P, Li L, et al. RAS/BRAF circulating tumor DNA mutations as a predictor of response to first-line chemotherapy in metastatic colorectal Cancer patients. Canadian Journal of Gastroenterology & Hepatology. 2018;**2018**:4248971

[86] Xu RH, Muro K, Morita S, et al. Modified XELIRI (capecitabine plus irinotecan) versus FOLFIRI (leucovorin, fluorouracil, and irinotecan), both either with or without bevacizumab, as second-line therapy for metastatic colorectal cancer (AXEPT): A multicentre, open-label, randomised, non-inferiority, phase 3 trial. The Lancet Oncology. 2018;**19**(5):660-671. DOI:10.1016/S1470-2045(18)30140-2

[87] Defachelles AS, Bogart E, Casanova M, Merks JHM, Bisogno G, Calareso G, et al. Randomized phase

II trial of vincristine-irinotecan with or without temozolomide, in children and adults with relapsed or refractory rhabdomyosarcoma: A European paediatric soft tissue sarcoma study group and innovative therapies for children with cancer trial. Journal of Clinical Oncology. 2021;**39**(27):2979-2990

[88] Koganemaru S, Kuboki Y, Koga Y, Kojima T, Yamauchi M, Maeda N, et al. U3-1402, a novel HER3-targeting antibody-drug conjugate, for the treatment of colorectal Cancer. Molecular Cancer Therapeutics. 2019;**18**(11):2043-2050

[89] Yao Z, Zheng W, Zhang X, Xiong H, Qian Y, Fan C. Hydroxycamptothecin prevents fibrotic pathways in fibroblasts In vitro. IUBMB Life. 2019;**71**(5):653-662

[90] Laco GS, Du W, Kohlhagen G, Sayer JM, Jerina DM, Burke TG, et al. Analysis of human topoisomerase I inhibition and interaction with the cleavage site +1 deoxyguanosine, via in vitro experiments and molecular modeling studies. Bioorganic & Medicinal Chemistry. 2004;**12**(19):5225-5235

[91] Borkar MR, Martis EAF, Nandan S, Patil RH, Shelar A, Iyer KR, et al. Identification of potential antileishmanial 1,3-disubstituted-4-hydroxy-6-methylpyridin-2(1H)-ones, in vitro metabolic stability, cytotoxicity and molecular modeling studies. Chemico-Biological Interactions. 2022 Jan;**5**(351):109758

[92] Kardile RA, Sarkate AP, Borude AS, Mane RS, Lokwani DK, Tiwari SV, et al. Design and synthesis of novel conformationally constrained 7,12-dihydrodibenzo[b,h][1,6] naphthyridine and 7H-Chromeno[3,2-c] quinoline derivatives as topoisomerase I inhibitors: In vitro screening, molecular docking and ADME predictions. Bioorganic Chemistry. 2021;**115**:105174

[93] de Souza PL, Cooper MR, Imondi AR, Myers CE. 9-Aminocamptothecin: A topoisomerase I inhibitor with preclinical activity in prostate cancer. Clinical Cancer Research. 1997;**3**(2):287-294

[94] Halder N, Dzhemileva LU, Ramazanov IR, D'yakonov VA, Dzhemilev UM, Rath H. Comparative in vitro studies of the topoisomerase I inhibition and anticancer activities of metallated N-confused porphyrins and metallated porphyrins. ChemMedChem. 2020;**15**(7):632-642

[95] Haleel A, Mahendiran D, Veena V, Sakthivel N, Rahiman AK. Antioxidant, DNA interaction, VEGFR2 kinase, topoisomerase I and in vitro cytotoxic activities of heteroleptic copper(II) complexes of tetrazolo[1,5-a]pyrimidines and diimines. Materials Science & Engineering. C, Materials for Biological Applications. 2016;**68**:366-382

[96] Parvez MM, Basit A, Jariwala PB, Gáborik Z, Kis E, Heyward S, et al. Quantitative investigation of irinotecan metabolism, transport, and gut microbiome activation. Drug Metabolism and Disposition. 2021;**49**(8):683-693

[97] Ogitani Y, Hagihara K, Oitate M, Naito H, Agatsuma T. Bystander killing effect of DS-8201a, a novel anti-human epidermal growth factor receptor 2 antibody-drug conjugate, in tumors with human epidermal growth factor receptor 2 heterogeneity. Cancer Science. 2016;**107**(7):1039-1046

[98] Fukuda T, Nanjo Y, Fujimoto M, Yoshida K, Natsui Y, Ishibashi F, et al. Lamellarin-inspired potent topoisomerase I inhibitors with

the unprecedented benzo[g][1] benzopyrano[4,3-b]indol-6(13H)-one scaffold. Bioorganic & Medicinal Chemistry. 2019;**27**(2):265-277

[99] Ki DH, Oppel F, Durbin AD, Look AT. Mechanisms underlying synergy between DNA topoisomerase I-targeted drugs and mTOR kinase inhibitors in NF1-associated malignant peripheral nerve sheath tumors. Oncogene. 2019;**38**(39):6585-6598

[100] Liu H, Lu H, Liao L, Zhang X, Gong T, Zhang Z. Lipid nanoparticles loaded with 7-ethyl-10-hydroxycamptothecin-phospholipid complex: in vitro and in vivo studies. Drug Delivery. 2015;**22**(6):701-709

[101] Lei CS, Hou YC, Pai MH, Lin MT, Yeh SL. Effects of quercetin combined with anticancer drugs on metastasis-associated factors of gastric cancer cells: in vitro and in vivo studies. The Journal of Nutritional Biochemistry. 2018;**51**:105-113

Chapter 5

# Marker Assisted Selection in Groundnut

*Diriba Beyene Goonde and Seltene Abady*


## Abstract

Groundnut (*Arachis hypogaea* L.) is an important oilseed crop worldwide. Objective of this review is to highlight molecular breeding approach such as marker assisted selection on groundnut improvement with future perspectives. The review analyzed application of marker assisted selection including simple sequence repeats, random amplified polymorphism DNAs, single nucleotide polymorphism, amplified fragment length polymorphism and inter simple sequence repeats on groundnut improvement. Among the molecular markers, random amplified polymorphic DNA is a rapid method for developing genetic maps and to determine DNA fragments to characterize peanut cultivars. DArTseq is used for SNP discovery and genotyping, which enables considerable discovery of SNPs in a wide variety of non-model organisms and provides measures of genetic divergence. Polymorphism screening performed using these newly developed SSRs will greatly increase the density of SSR markers in the peanut genetic map in the future.


**Keywords:** genome, molecular markers, simple sequence repeats, groundnut, marker assisted selection

## 1. Introduction

Groundnut (*Arachis hypogaea* L.), also known as peanut, is a member of genus Arachis and family Leguminosae [1]. Peanuts are key oilseed and food-legume crops for both humans and livestock in tropical and subtropical regions, and globally they are the fourth largest source of edible oil and third most important source vegetable protein. Its seed contain about 50% of edible oil and the remaining 50% of the seed has high quality protein (36.4%), carbohydrate in the range (6–24.9%), minerals and vitamin [2]. It is believed to have originated in the southern Bolivia to northern Argentina region of South America. Cultivated Groundnut (A. hypogaea L., 2n = 4x = 40, AABB) is self-pollinating allotetraploid legume crop belonging to the Fabaceae family [3].

Groundnut was introduced to Ethiopia by Italian explorers in the 1920s [4]. Globally China ranks first in groundnut production with 17.39 million tonnes followed by India 6.95 million tonnes, Nigeria 2.88 million tonnes, Sudan 2.88 million tonnes and Ethiopia ranks 31th with 0.129 million tonnes with national mean yield is 1.75 tons/ha, and the total area under groundnut production is 115,291 ha [5]. The most common groundnut production constraint in Ethiopia in general and the southern region, in particular, were the lack of access to improved seeds, biotic, abiotic stress, and the use of low-yielding local varieties [6, 7]. Therefore, the objective of this review is to highlight molecular breeding approaches such as marker assisted selection on groundnut improvement and opportunities, challenges with future perespectives of the crop.

## 2. Status production of groundnut

Groundnut are predominantly grown in developing countries (Africa and Asia where the crop finds appropriate climate for optimum production). Although, the production is concentrated in Asia (50% global area and 68% global production) and Africa (46% global area and 24% of global production) (**Table 1**).

| Country | Production (tons in million) | Hectare (tons in million) | Yield (kg/hectare) |
|---|---|---|---|
| China | 16.685 | 4.541 | 3.674 |
| India | 6.857 | 5.800 | 1.182 |
| Nigeria | 3.028 | 2.680 | 1.130 |
| USA | 2.578 | 0.626 | 4.118 |
| Sudan | 1.826 | 2.315 | 788.8 |
| Ethiopia (31th) | 0.129 | 74.861 | 1.731 |

*Source: [5].*

**Table 1.**
*The top leading country in production and productivity of groundnut in the world.*

## 3. Genomic resources

Genomic resources such as molecular markers are powerful tools to characterize and harness the genetic variation present in the germplasm collection. Peanut has comparatively lower genomic resources (including transcriptome data) compared to other legumes like medicago, lotus and chickpea [8], robust molecular markers, specifically the genetic ones, as they provide the insight into the functional information. However, the available peanut high throughput transcriptome sequences are not complete; many have low N50 values, ranging from 500 to 750 bp [9]. Because peanut

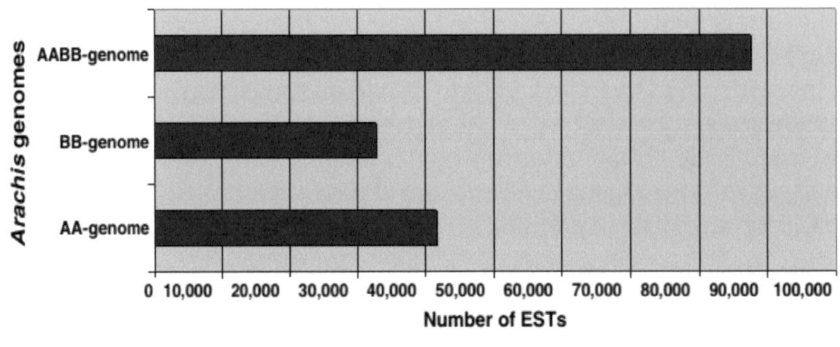


**Figure 1.**
*Publicly available expressed sequence tags in different Arachis species.* Source: *[9].*

has such a large number of genes, it is important to have a good representation of the transcriptome. On other hand, ESTs data of cultivated peanut is still remain unexplored for the development of SSR markers (**Figure 1**).

## 4. Marker assisted selection

Marker-assisted selection has been a plant breeding tool since it was proposed by Sax in 1923 [10]. The theory behind this method is that plant breeders could observe easy-to-score phenotypes to select difficult-to-score or low heritability traits that are linked to them. Marker assisted selection is the indirect selection of selected or desired plant phenotype depending on the closely linked DNA marker. MAS is an efficient molecular tool for breeding, in which markers linked with the desired genes are used for indirect selection for that gene in non-segregating or segregating populations. MAS is an important method for the selection of traits that are difficult, like, biotic and abiotic stress tolerance in a crop [11]. Compared with conventional phenotypic selection, MAS is not influenced by environmental conditions because it detects the structural polymorphisms at the molecular level. Further MAS is cheaper and less labour intensive, allows selection in off-season nurseries and has a potential to accelerate the breeding process [12].

Molecular markers among all genomic resources, molecular markers have direct use for germplasm characterization, trait mapping and molecular breeding. Several marker systems have been developed during the last three decades. For instance restriction fragment length polymorphisms (RFLPs), random amplified polymorphic DNAs (RAPDs), amplified fragment length polymorphisms (AFLPs) and diversity arrays technology (DArT) markers have proved their utility from time to time [13]. However, simple sequence repeats (SSRs) or microsatellites and single nucleotide polymorphism (SNP) markers are generally preferred for plant genetics and breeding

| Molecular markers | Type | Amplification of markers/technique used for identification |
|---|---|---|
| Restriction fragments length polymorphism | Co-dominant | Depends on point of mutation |
| Sequence characterized amplified region | Co-dominant | Depend on mutation at primer annealing site in the specific region of DNA strand |
| Amplified fragment length polymorphism | Dominant | Depend on mutation at primer annealing site in the target DNA and change restriction site in the target DNA |
| Simple Sequence repeats | Co-dominant | Difference in the number of repeats of motif |
| Diversity arrays technology | Dominant | Micro array hybridization, they produced from genomic libraries through amplification of candidate or random clones |
| Single nucleotide polymorphism | Co-dominant | Point mutation along with sequence information |
| Sequence characterized region | Co-dominant | Depend on mutation at primer annealing site in specific region of DNA strand |

**Table 2.**
*Commonly used molecular markers are as below.*

applications. While SSR markers are multi-allelic, codominant and easy to use, the SNP markers are highly amenable to high-throughput genotyping approaches. Development and application of SNP markers, however, is still not routine in crop species and especially not in low-tech laboratories (**Table 2**).

### 4.1 Markers for target traits

The approach of identifying markers for targets traits swiftly changed with the development of linkage maps in groundnut [14]. Seed weight is controlled by a combination seed features such as Seed length, Seed width, and seed thickness. Several genes for seed-related traits have been obtained in many crops using the forward genetic strategy and reverse genetic strategies [15]. QTL analysis was used for identification of QTLs for several important traits such as drought tolerance related traits, resistance to foliar disease and nutritional quality traits (**Table 3**).

### 4.2 Molecular markers in groundnut

Cultivated groundnut has been analyzed by several markers systems including RFLPs, RAPDs, AFLPs and SSRs.

#### *4.2.1 Restriction fragment length polymorphism*

Restriction Fragment Length Polymorphism (RFLPs) represented the first marker system that had a large number of polymorphisms. They are used widely both to create linkage maps and to implement indirect selection strategies. In A. hypogaea, little molecular variation has been detected by using RFLP technologies have been used to analyze species in section Arachis (representing taxa that will hybridize with

| Population | Traits | Marker |
|---|---|---|
| Yuanza 9102 × ICGV 86699 | Rust resistance | AFLP |
| TAG 24 × GPBD 4 | Rust resistance | SSR |
| TAG 24 × GPBD 4, TG 26 × GPBD 4 | LLS and rust resistance | SSR |
| Zhonghua 5 × J 11 | Aflatoxin contamination | AFLP |
| TAG 24 × ICGV 86031 | Drought tolerance | SSR |
| Tamrun OL01 × BSS 56 | Pod and kernel traits | SSR |
| TG 26 × GPBD 4 | Protein content | SSR |
| TG 26 × GPBD 4 | oil content, and oil quality | SSR |
| Germplasm accessions and breeding lines | High oleic acid content (FAD2A) | Real time-PCR |
| Arachis hypogaea × TxAg-7 M | arenaria resistance | RFLP |
| Yuanza 9102 × Chico | Bacterial wilt resistance | SSR |

*RAPD, Randomly Amplified Polymorphic DNA; RFLP, Restriction Fragment Length Polymorphism; SSR, Simple Sequence Repeat; AFLP, Amplified Fragment Length Polymorphism; AS-PCR, Allele Specific Polymerase Chain Reaction. #FAD, Fatty acid desaturase;; ELS, Early Leaf Spot; LLS, Late Leaf Spot.*

**Table 3.**
*Molecular markers associated with trait specific genes/QTLs in groundnut.*

A. hypogaea) and clusters that formed using multivariate analyses [16] correspond closely with morphological groups [17]; tetraploids were clearly separated from diploids in both investigations. Stalker *et al.* [18] utilized RFLPs to examine genetic diversity among 18 accessions of A. duranensis Krapov. and We. Gregory and found a large amount of variation in the species. Individual accessions also could be uniquely identified by RFLP patterns. Kochert et al. [19] concluded that the cultivated peanut resulted from a cross between A. duranensis and A. ipaensis Krapov. and W.C. Gregory, and chloroplast analysis indicated that A. duranensis was the female progenitor of the cross. An RFLP map was developed for peanut by analyzing an F2 population from the diploid (2n = 2x = 20) interspecific cross of A. stenosperma Krapov. and W.C. Gregory (ace, HLK 410) and A. cardenasii Krapov.and W.C. Gregory (ace, GKP 10017). The linkage map covered 1063 cM with 117 markers in 11 linkage groups [20]. Fifteen unassociated markers also were reported. A second molecular map of peanut was created by Burow, Patterson, and Simpson using the tetraploid cross Florunner X 4x [A. batizocoi Krapov. and W.C. Gregory (A. cardenasii X A. diogoi Hoehne)] Burow (pers. commun.). Most of the 380 RFLP markers that have been mapped had disomic inheritance, with the exception of one linkage group which may be polysomic.

#### *4.2.2 Simple sequence repeats*

Simple sequence repeats (SSRs) are genomic fragments that consist of tandemly repeated units that are present in both coding and non-coding regions of the genome [21]. SSR markers, designed by flanking sequences, are useful for and widely applied in plant genetic analyses and marker-assisted selection breeding. Currently, g-SSR markers are common and popular for such analyses, and they have wide applications in molecular genetics and breeding, because they have multiple advantages, such as simplicity, abundance, ubiquity, variation, co-dominance, and multi-allelism [21]. Even though the peanut genome had not yet been resolved. With the recent completion of genome sequencing of peanut and two diploid progenitor species, A. duranensis and A. ipaensis, a large number of genome-wide g-SSRs were identified. SSR markers linked to resistance to early leaf spot, groundnut rosette disease, and rust and aflatoxin contamination across African cultivated groundnut varieties were identified useful to identify suitable parents for mapping populations or breeding [22].

Genotypes with similar genetic backgrounds tended to cluster in the same subgroup, indicating the effectiveness of SNP markers in assigning the tested genotypes into homogenous groups [23].

Simple sequence repeat (SSR) alleles associated with agronomic traits in at least two environments. These markers were further investigated for their potential use in genetic studies by ascertaining their genetic diversity in the natural population.

#### *4.2.3 Inter simple sequence repeats*

Inter Simple Sequence Repeats (ISSR) marker has been reported as a rapid, reproducible, and cheap fingerprinting technique based on the variation found in the regions between microsatellites. It is a fast, inexpensive genotyping technique based on variation in the regions between microsatellites (Inter Simple Sequence Repeats analyses offer breeders and geneticists with competent means to link phenotypic and genotypic variations in various fields of plant improvement) [24].

### *4.2.4 Randomly amplified polymorphic DNA markers*

Among the molecular markers, random amplified polymorphic DNA (RAPD) is a rapid method for developing genetic maps and to determine DNA fragments to characterize peanut cultivars. PCR based Randomly Amplified Polymorphic DNA markers are good genetic markers because they give rapid results, economically convenient and use small oligonucleotide primers. With a small quantity of template, a very large number of fragments are generated from different regions of the genome and hence, multiple loci may be examined very quickly [25, 26].

### *4.2.5 Diversity arrays technology*

Diversity Arrays Technology (DArT), which is based on genome complexity reduction and SNP detection through hybridization of PCR fragments has been used in genome-wide association studies (GWAS), construction of dense linkage maps and mapping quantitative trait loci (**Table 4**) [27].

SSR, simple sequence repeat markers, TEM, Transposable element markers, RAPD, random amplified polymorphic DNA.

| Marker name | Marker type | Marker sequence | |
|---|---|---|---|
| | | Forward primer | Reverse primer |
| IPAHM103 | SSR | GCATTCACCACCATAGTCCA | TCCTCTGACTTTCCTCCATCA |
| GM1536 | SSR | AAAGCCCTGAAAAGAAAGCAG | ATGCATTTGCAGGTTCTGGT |
| GM2301 | SSR | GTAACCACAGCTGGCATGAAC | CTTCAAGAACCCACCAACAC |
| GM2079 | SSR | GGCCAAGGAGAAGAAGAAAGA | GAAGGAGTAGTGGTGCTGCTG |
| GM1991 | SSR | GAAAATGATGCCGAGAAATGT | GGGGAGAGATGCAGAAAGAGA |
| TE 360 | TEM | GGATATGATGCCCATAGCTGA | TGCTGACTACTTGCAATGCC |
| TE 498 | TEM | ATGACTTACATGTAGCAATTG | TGAAAGGAGTCAAAGGTCATG |
| S197 | RAPD | CTGTCGAACCATGGAAGAAGATCC | CCAACTTGATGGTAGAAGTATGCTT |
| AHCW0061 | SSR | TCATGTGAATTTGTGGACGGT | CCAGGTTTTTGAGGTCCCTGA |
| AHCW0310 | SSR | GTTCAAGGGCTGTGCATTGG | GGGTTCGACTCCCGTCTTAT |
| AHCW0545 | SSR | ACAGAAGAAGAAACAGCGCG | TTCCGTCATGTGCTTCGGAA |
| AHCW0618 | SSR | AAATTTGAGCACGCATCCCC | TGTCTTTTTCCTCGCCTTTGT |
| AHCW0700 | SSR | TGGAAGTTTCACGGGACAGG | GTAGCAAGCTTCCCCACCAT |
| AHCW0768 | SSR | GGACCCATTTTTGCAAGAGAGA | CGGATTGCAACATTGGCGAA |
| AHCW1250 | SSR | ACAGCTGCCTCTTCTCTGTG | CCCACTCAAAATCGGATTTGGA |
| AHCW1510 | SSR | TCCTGCACCATGACCATGAA | TGTTCGGCACCAATCTGTCA |
| AHCW1765 | SSR | CGCTGGTCTGGCATTTAACG | AAGGGAGGAGGAGTTGGGTT |
| AHCW1862 | SSR | TGTTCAGGGATGTGTTTGGACT | GGGCAAGCTCTTTAAACTGCA |

**Table 4.**
*Some molecular marker systems developed for genetic analysis and breeding in groundnut.* Source: *[6, 7]*.

## 5. Perspectives

Molecular markers can assist in the selection process with phenotypic selection and speed up the pace of breeding cycle, in recent times technologies such as next generation sequencing i.e. low with high throughput, Genome Selection and Genotype by sequencing can be used to achieve the desired goal in molecular breeding approaches. The genome/gene space sequence would provide the opportunities to link the phenotype with genes. The future of peanut genomics and use of molecular tools in breeding seems to be bright that will ensure the peanut improvement for different production as well as quality constraints.

## 6. Conclusion

Molecular markers are used to identify quantitative trait loci which enhance the efficiency of selecting complex trait in plant breeding. MAS is an efficient molecular tool for breeding, in which markers linked with the desired genes are used for indirect selection for that gene in non-segregating or segregating populations. Now a days, DArT, SSR, SNP, ISSR, etc. with high throughput technologies are very exciting markers, which enhances the crop with desired traits and induces tolerance against biotic and abiotic stresses in a short period of time.

Molecular markers can provide information that can help define the distinctiveness of species and their ranking according to the number of close relatives and their ranking according to the phylogenetic position. RAPD patterns generated from peanut cultivars could be used as genomic fingerprint to establish the identity of a given genotype. The utilization of DArT marker system may limit efficient genetic analysis of groundnut genetic resources for cultivar development. Development of highly discriminative and informative DArT markers is useful for genetic analysis and breeding in groundnut.

A desirable molecular marker should have high polymorphism, frequent occurrence, should be easy to use and should be quick, co-dominant inheritance, equally dispersed all over the genome, high transferability and reproducibility, less expensive and phenotypically neutral. However, it will still take some time before cost-effective SNP genotyping platforms are available for genotyping the tetraploid peanut germplasm collections or peanut mapping populations. Extension of SNP-based maps to the tetraploid has not been accomplished yet, and will require separation of A- and B-genome sequences, but is expected to greatly accelerate genetic mapping and marker-assisted selection.

## Acknowledgements

Wallaga University and Haramaya University Fully acknowledged for giving chance to Fellow Doctor of Philosophy.

Seltene Abady (PhD) is sincerely acknowledged for guiding me the review.

## Author details

Diriba Beyene Goonde[1,2]* and Seltene Abady[2]

1 Department of Philosophy, Research and Technology, Research and Technology Park, Wallaga University, Ethiopia

2 Haramaya University, Ethiopia

*Address all correspondence to: hordofabeyene2018@gmail.com

## References

[1] Krapovickas A and Gregory WC. 2007. Taxonomy of the Genus Arachis (Leguminosae). Translated by DE Williams, CE Simpson. Bonplandia. 16:1-205

[2] Baraker B, Jha SK, Wani SP, Garg KK. Effect of balanced fertilizer management practices on factor of productivity on Groundnut (*Arachis hypogaea* L.) cultivation. International Journal of Chemical Studies. 2017;**5**(4):1288-1291

[3] Janila P, Nigam SN, Pandey MK, Nagesh P, Varshney RK. Groundnut improvement: Use of genetic and genomic tools. Frontiers in Plant Science. 2013;**4**:1-16

[4] Wakjira A. A review of the recent groundnut breeding activities in Ethiopia. In: Paper Presented at the First National Oilseeds Work Shop, 3-5 Dec. Addis Ababa, Ethiopia. 1991

[5] FAO. Rome, Italy: Food and Agriculture Organization of the United Nations; 2021

[6] Abady S, Shimelis H, Janila P. Farmers' perceived constraints to groundnut production, their variety choice and preferred traits in eastern Ethiopia: Implications for drought-tolerance breeding. Journal of Crop Improvement. 2019a;**33**:1-17. DOI: 10.1080/154275 28.2019. 1625836

[7] Abady S, Shimelis H, Janila P, Mashilo J. Groundnut (Arachishypogaea L.) improvement in sub-Saharan Africa: A review. Acta Agriculturae Scandinavica, Section B-Soil & Plant Science. 2019;**69**(6):528-545. DOI: 10.1080/09064710.2019.1601252

[8] Sato S, Nakamura Y, Kaneko T, Asamizu E, Kato T, Nakao M, et al. Genome structure of the legume, *Lotus japonicus*. DNA Research. 2008;**15**:227-239

[9] Guimaraes PM et al. Global transcriptome analysis of two wild relatives of peanut under drought and fungi infection. BMC Genomics. 2012;**13**:387

[10] Arus P, Moreno-Gonzalez J. Marker-assisted selection. In: Hayward MD, Bosemark NO, Romagosa I, editors. Plant Breeding: Principles and Prospects. London: Champman and Hall; 1993. pp. 314-331

[11] Das G, Patra JK, Baek KH. Insight into MAS: 2017 a molecular tool for development of stress resistant and quality of rice through gene stacking. Frontiers in Plant Science. 2017;**8**:985

[12] Kumpatla SP, Buyyarapu R, Abdurakhmonov IY, Mammadov JA. Genomics-assisted plant breeding in the 21st Century. In: Abdurakhmonov I, editor. Technological Advances and Progress. In Tech; 2012. pp. 132-184

[13] Varshney RK, Hoisington DA, Tyagi AK. Advances in cereal genomics and applications in crop breeding. Trends in Biotechnology. 2006;**24**:490-499

[14] Pandey M et al. Advances in Arachis genomics for peanut improvement. Biotechnology Advances. 2012;**30**:639-651

[15] Si L, Chen J, Huang X, Gong H, Luo J, Hou Q, et al. OsSPL13 controls grain size in cultivated rice. Nature Genetics. 2016;**48**:447-456. DOI: 10.1038/ng.3518

[16] Kochert G, Halward TM, Branch WD, Simpson CE. RFLP variability in peanut cultivars and wild species. Theoretical and Applied Genetics. 1991;**81**:565-570

[17] Stalker HT. A morphological appraisal of wild species in section

Arachis of peanuts. Peanut Science. 1990;**17**:117-122

[18] Stalker HT, Simpson CE. Genetic resources in Arachis. In: Pattee HE, Stalker HT, editors. Advances in Peanut Science. American Peanut Research and Education Society. Stillwater. 1995. pp. 14-53

[19] Kochert G, Stalker HT, Ginenes M, Galgaro L, Moore K. RFLP and cytogenetic evidence for the progenitor species of allotetraploid cultivated peanut, Arachis hypogaea (Leguminosae). American Journal of Botany. 1996;**83**:1282-1291

[20] Halward TM, Stalker HT, Kochert G. Development of an RFLP linkage map in diploid peanut species. Theoretical and Applied Genetics. 1993;**87**(3):379-384

[21] Haq SU, Jain R, Sharma M, Kachhwaha S, Kothari SL. Identification and characterization of microsatellites in expressed sequence tags and their cross transferability in different plants. International Journal of Genomics. 2014;**2014**:863948

[22] Kanyika BTN, Lungu D, Mweetwa AM, Kaimoyo E, Njung'e VM, Monyo ES, et al. Identification of groundnut (Arachis hypogaea) SSR markers suitable for multiple resistance traits QTL mapping in African germplasm. Electronic Journal of Biotechnology. 2015;**18**:61-67

[23] Adu BG, Badu-Apraku B, Akromah R, Garcia-Oliveira AL, Awuku FJ, Gedil M. Genetic diversity and population structure of early-maturing tropical maize inbred lines 146 using SNP markers. PLoS One. 2019;**14**(4):e0214810. DOI: 10.1371/journal. Pone. 0214810

[24] Tadele S, Mekbib F, Tesfaye K. Genetic diversity of coffee (Coffea arabica L.) landraces from Southern Ethiopia as revealed by inter simple sequence repeat marker. Global Advanced Research. Journal of Agricultural Science. 2014;**3**(1):024-034

[25] Edae EA, Byrne PF, Haley SD, Lopes MS, Reynolds MP. Genome-wide association mapping of yield and yield components of spring wheat under contrasting moisture regimes. Theory Applied Genetics. 2014;**127**:791-807

[26] Kumar A, Jain S, Elias EM, Ibrahim M, Sharma LK. An overview of QTL identification and marker-assisted selection for grain protein content in wheat. In: Eco-friendly Agro-biological Techniques for Enhancing Crop Productivity. Singapore: Springer; 2018. pp. 245-274

[27] Abu Zaitoun SY, Jamous RM, Shtaya MJ, Mallah OB, Eid IS, Ali-Shtayeh MS. Characterizing palestinian snake melon (Cucumis Melo Var. Flexuosus) germplasm diversity and structure using SNP and DArTseq markers. BMC Plant Biology. 2018;**18**:246